ERIC A LORD, ALAN L MACKAY AND
S RANGANATHAN

New Geometries for New Materials

Recent advances in materials science have given rise to novel materials with unique properties, through the manipulation of structure at the atomic level. Elucidating the shape and form of matter at this scale requires the application of mathematical concepts. This book presents the geometrical ideas that are being developed and integrated into materials science to provide descriptors and enable visualisation of the atomic arrangements in three-dimensional (3D) space. Emphasis is placed on the intuitive understanding of geometrical principles, presented through numerous illustrations. Mathematical complexity is kept to a minimum and only a superficial knowledge of vectors and matrices is required, making this an accessible introduction to the area. With a comprehensive reference list, this book will appeal especially to those working in crystallography, solid state and materials science.

DR ERIC A. LORD is a Visiting Scientist in the Department of Metallurgy at the Indian Institute of Science.

PROF. ALAN L. MACKAY retired in 1991 as a Professor Emeritus in the School of Crystallography at the University of London; however, he continues to pursue several research collaborations. He is a Fellow of the Royal Society.

PROF. S. RANGANATHAN is an Honorary Professor in the Department of Metallurgy at the Indian Institute of Sciences and has contributed to some 250 papers in scientific journals.

NEW GEOMETRIES FOR NEW MATERIALS

ERIC A LORD
Department of Metallurgy, Indian Institute of Science, Bangalore, India

ALAN L MACKAY
School of Crystallography, Birkbeck College, University of London, London, UK

S RANGANATHAN
Department of Metallurgy, Indian Institute of Science, Bangalore, India

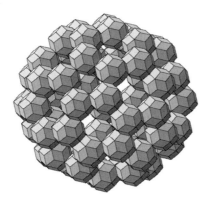

CAMBRIDGE
UNIVERSITY PRESS

CAMBRIDGE UNIVERSITY PRESS
Cambridge, New York, Melbourne, Madrid, Cape Town, Singapore, São Paulo

Cambridge University Press
The Edinburgh Building, Cambridge CB2 2RU, UK
Published in the United States of America by Cambridge University Press, New York

www.cambridge.org
Information on this title: www.cambridge.org/9780521861045

First published 2006

Printed in the United Kingdom at the University Press, Cambridge

A Catalogue record for this book is available from the British Library

Library of Congress Cataloguing in Publication data

ISBN-13 978-0-521-86104-5 hardback
ISBN-10 0-521-86104-7 hardback

Contents

Preface

Over the past few decades unprecedented and far-reaching discoveries have been taking place in the materials sciences – in solid state physics, crystallography, metallurgy, nanotechnology, microbiology… These discoveries have given rise to new materials with unusual and valuable properties and have led to a deeper understanding of how nature works – of how atoms combine to build the world. At the most basic level the study of the kinds of patterns and structures that can arise from the combination of subunits in three-dimensional (3D) space reveals a metastructure of underlying general principles. We are essentially dealing here with a *language*, the language of shape and form, the language of the geometry of 3D space.

The range of shapes and patterns that are possible in 3D space is independent of scale. Thus, for example, the C_{60} molecule has been named 'Buckminsterfullerene' or, colloquially, the 'Bucky ball' because the uniform spherical arrangement of its 60 atoms corresponds to the icosahedral geometry underlying Buckminster Fuller's geodesic dome constructions. Only the scale is different. Thus, though the majority of the structures that we have chosen to describe and to exemplify general principles are taken mainly from the literature of the materials sciences, and though it is to the materials scientist that our work is principally addressed, it is our hope that our presentation will not be without interest to a wider readership.

In seeking inroads into the problem of understanding how complex structures arise in nature, some of the more exotic geometries have been employed as an aid to thinking about structures. For example, much of the theoretical investigation of quasicrystals has made forays into 6D space; Stephen Hyde's work has shown how non-Euclidean geometry can throw light on the structure of networks; the real and hypothetical materials investigated by Terrones and his co-workers bring in the geometry of curved surfaces, and so on. As materials science has become increasingly mathematically oriented, mathematicians have in turn been stimulated to new investigations by discoveries in materials science. An exciting dialogue is emerging.

The field is vast, and growing. Clearly, a review such as this needs to be select-ive. What makes the study of structure fascinating is its appeal to the imagination. We have placed the emphasis on the intuitive feeling for 3D shapes and structures. That is the foundation on which any such study is built. We have, accordingly, attempted to keep mathematical details to a minimum. In the places where a math-ematical technique or demonstration seemed unavoidable we hope the uninitiated reader will bear with us. Extensive citations of our sources will, we trust, be help-ful for the curious reader who wishes or needs to pursue further some of the more esoteric aspects of our theme.

We have aimed to bring out clearly the interconnections among the topics dealt with in the various chapters – to emphasise tha unity of our subject matter. Hence, the reader will sometimes find the same structures occurring in different contexts, described from different viewpoints.

The writing of this book was undertaken as part of the project 'New Geometries for New Materials' sponsored by the Defence Research and Development Organi-sation, Ministry of Defense, Government of India (project reference DRDO/MMT/ SRG/526). Their support and encouragement is gratefully acknowledged. SR is grateful to the Homi Bhabha Fellowship Council for support. Some additional material related to this project can be found on our website http://met.iisc.ernet .in/~lord. Almost all the figures (not just those in Chapter 9) were produced using Ken Brakke's remarkable software Surface Evolver. We are indebted to him for making Surface Evolver freely available for downloading from the Internet.

Eric A. Lord
Alan L. Mackay
S. Ranganathan

1

Introduction

1.1 Atoms

The virtually infinite variety of phenomena – the properties of matter, living and non-living, its behaviour and transformations – are manifestations of the various ways in which a limited number of structural units – *atoms* – can combine with each other to build up more intricate structures. This is now common knowledge. When it was proposed by Democritus and Epicurus in the fourth century BC, it was an astonishing hypothesis. A definitive exposition of the Democritean atomic hypothesis, *De Rerum Natura*, was written by Lucretius (*c*. 60 BC) (English translation by R. E. Latham: Lucretius (1994)). Through the medium of an epic Latin poem, Lucretius demonstrates how numerous familiar phenomena can be rationally accounted for on the assumption that 'there exists only atoms and empty space'. Its intuitive insights and uncanny premonitions of modern physics are all the more amazing in view of the fact that experimentation was of no great interest to editors of the Greek philosophers – the arguments are based on thoughtful observation of familiar things. The recent elucidation of the anti-Kythera mechanism, a mechanical device from the time of Lucretius like an orrery, for predicting the positions of celestial objects, which could perhaps be used for determining the longitude, shows that the experimental tradition of Archimedes continued.

Lucretius, in criticising the theory of Anaxagoras, even describes what we might now call 'fractals': 'in speaking of the *homoeomeria* of things Anaxagoras means that the bones are formed of minute miniature bones . . . gold consists of grains of gold . . . fires of fires . . .'

Concerning the number of elements, Lucretius wisely refused to enter into unfounded metaphysical speculations, saying only that the number of different kinds of atom is finite. The traditionally held view, opposed by Lucretius, was that there are just *four* elements: earth, fire, air and water, which were supposed to correspond symbolically to the four regular polyhedra, cube, tetrahedron, octahedron

Figure 1.1 The five Platonic solids and the four elements; from Kepler's *Harmonice Mundi* (Kepler 1619).

and icosahedron (Figure 1.1). (The fifth regular polyhedron, the pentagonal dodecahedron, corresponded to the mysterious 'fifth essence' – the notion underlies the etymology of *quintessential*.)

In Plato's view, the correspondence between the four elements and four regular polyhedra was not only symbolic – he proposed that these polyhedra corresponded to the actual *shapes* of the atoms. For this reason the regular polyhedra are referred to as the 'Platonic solids'. Even the strangest ideas can contain a grain of truth: the regular and the semi-regular polyhedra ('the 5 Platonic and 13 Archimedean solids') are indeed prominent features in the microstructure of solids and liquids, but in the shapes of *clusters* of atoms rather than of individual atoms. Plato, in his 'Timaeus', sees the polyhedra as themselves built of 45-90-45 degree and 60-90-30 degree triangles, which are just the figures still to be found in a school geometry set. The five symmetrical 'Platonic' configurations were already known a millenium before the time of Plato. Over 400 carved stone balls from neolithic times have been found at various sites in northern Scotland. The Universities of Aberdeen and Glasgow possess extensive collections of these objects. A particularly fine set of five,

Figure 1.2 Scottish neolithic stone carvings corresponding to the regular polyhedra, in the Ashmolean Museum, Oxford. Their purpose is unknown. Photo by Graham Challifour for Critchlow (1979). Reproduced by kind permission of Graham Challifour and Keith Critchlow.

exhibiting the five configurations, is owned by the Ashmolean Museum, Oxford (see Figure 1.2).

The notion of the *four elements* had a powerful hold over scientific thinking until surprisingly recent times. Joseph Priestley discovered the element oxygen in 1774. Antoine Lavoisier repeated Priestley's experiments and understood their significance: air is composed of several gases, and oxygen is one of them. He also suggested that water, too, is a compound. These insights marked the beginnings of modern chemistry. This was the response of Antoine Baumé (1728–1804), the speaker of the Paris Academy of Science: 'The elements or base components of bodies have been recognised and determined by physicists of every century and every nation. It is inadmissible that the elements recognised for 2000 years should now be included in the category of compound substances. They have served as the basis of discoveries and theories . . . We should deprive these discoveries of all credibility if fire, water, air and earth were no longer to count as elements.'

This epitomises the fact that human beings, like crystals, are creatures of habit. Ways of approaching problems, once they have proved successful, tend to become rigidified into traditions. The history of science is replete with examples of how well-established patterns of thought have hampered the emergence of new developments that, once they have emerged, open up new worlds for exploration.

1.2 Geometry

For more than 2000 years mathematicians took it for granted that geometry, as systematised and presented by Euclid, was the only possible geometry, until the discovery by Bolyai and Lobachevski in the nineteenth century of 'non-Euclidean geometry'. The geometry of Bolyai and Lobachevski is the geometry of the hyperbolic plane, H_2, which readily generalises to *hyperbolic spaces*, H_n. The hypersurfaces S_n of hyperspheres in Euclidean spaces E_{n+1} give another family of non-Euclidean geometries. (The subscripts indicate the dimensions of the space in question). Spherical trigonometry – geometry on the surface of a sphere – is the geometry of S_2; it was investigated by Hipparchus (150 BC) and the basic theorems are in the *Spheraeca* of Menelaus and in Ptolemy's *Almagest*. Two thousand years passed before anyone noticed that this is a *non-Euclidean* geom-etry of two dimensions! The spaces E_n, S_n and H_n are spaces of constant Gaussian curvature (zero, positive and negative, respectively). The formulae of hyperbolic geometry H_2 are just those of the geometry on the surface of a sphere, but with the radius of the latter set to i, the square root of minus one. Riemann generalised still further, developing the geometry of spaces in which the metrical properties varied continuously from point to point – a generalisation to higher dimensions of the intrinsic metrical properties of surfaces, which had been investigated by Gauss. The most general – and in a sense the most primitive – kind of geometry is topology, which takes no account of metrical properties; it deals only with the continuity and combinatorial properties of geometrical figures.

Although material structures, of course, exist in three-dimensional (3D) Euclidean space, some of the more exotic geometrical concepts have recently entered into the materials sciences, providing new and stimulating ways of thinking about these structures. We shall occasionally touch upon some of these developments, which serve to indicate the increasingly important role of mathematics in the science of materials.

1.3 Crystallography

A major mathematical contribution to our present understanding of the atomic constitution of crystalline solids was the work of Schoenflies, Fedorov and Barlow which classified triply periodic patterns in E_3 – there are just 230 different possible types, characterised by their symmetries. The geometrical theory of the symmetries of all possible crystals is one of the triumphs of nineteenth century mathematics. It is an elaborate edifice built on the basis of a simple assumption, namely that an 'ideal' crystal consists of an infinite number of identical units arrayed in space so that all have identical surroundings.

The experimental verification came with the introduction of X-ray structure analysis. The diffraction of X-rays by crystals was discovered by Max von Laue in 1912. He received a Nobel prize for this discovery two years later. Lawrence Bragg, the developer of the method along with his father William Bragg, revealed the atomic structure of sodium chloride as an array of alternating sodium and chlorine ions like a three-dimensional chessboard. Lawrence Bragg heard the news of his Nobel prize in 1916 while serving as an officer on the Western Front. With their technique they were able, by 1923, to laboriously produce pictures of the atoms of calcium, silicon and oxygen arranged in the mineral diopside. Since then, the arrangement of atoms in all matter, living and non-living, has been the basis of our understanding of its properties and behaviour.

Along with the development of X-ray diffraction techniques, the 230 space groups became the key to understanding crystalline structures. A picture emerged of a 'perfect' solid consisting of an arrangement of atoms 'decorating' the interior of a unit cell, which, repeated by translation, produces the whole structure. Unfortunately, this elegant scheme for a long time had a constraining effect on crystallographers somewhat analogous to the effect of Euclid's scheme on geometers. It became the paradigm. Important features of real materials were called 'defects' and materials that did not fit the scheme were dismissed as 'disordered'.

1.4 Generalised Crystallography

The discovery of quasicrystals has brought home forcefully to crystallographers the fact that a material structure could be highly ordered without being periodic. There are other, perhaps more interesting, ways in which structures can be orderly and systematic. Atoms and molecules know nothing of unit cells and they know no group theory; they simply respond to their immediate collective environment. Triple periodicity, when it arises, is a necessary consequence – an epiphenomenon – of more localised ordering principles. (An amusing illustration of the dominance of old ways of thinking is the modelling of quasicrystal structure in terms of a *pair* of 'unit cells' instead of one!)

A detailed understanding of how large-scale order (of any kind, not necessarily periodicity) arises from local ordering principles remains elusive. A theorem, due to Boris Nikolaievich Delone (Delone *et al.* 1934; 1976) throws some light on the way in which periodicity can arise from purely local conditions. A *Delone set* (r, R) in E_n is a set of points with the property that every sphere of radius r contains at most one of the points and every sphere of radius R contains at least one of the points (i.e. the points can be thought of as centres of hard spheres of radius r, and there are no large voids in the arrangement). Suppose, further, that for some length ρ, the configurations within spheres of radius $\rho + 2R$ centered at the points of the set are all congruent, and that the symmetry of this configuration is the same as the symmetry of the configuration within a radius ρ. Then *the point set is* (n-tuply) *periodic*. A simple proof of this assertion has been given by Senechal (1986). In E_2, ρ can be taken to be $4R$ and in E_3 it is conjectured that $\rho = 6R$.

The proposal that the scope of theoretical crystallography could, and should, be extended to embrace the study of systematic structures more general than the classical triply periodic structures of 'perfect' crystals, has long been advocated

(Bernal & Carlisle 1968; Mackay 1975). For a major portion of the twentieth century experimental crystallography was dominated by X-ray diffraction – a circumstance that restricted theoretical crystallography mainly to those structures that could be understood and described in terms of lattices and their reciprocal lattices. X-ray diffraction detected and emphasised the periodicities of a structure. In recent decades new experimental and observational techniques have become available, especially high-resolution electron microscopy, and have made 'generalised crystallography' a real possibility. Crystal structure analysis using the scattering of X-rays by crystals of the material under investigation has been such an enormously successful technique that other methods have been eclipsed. The technique has been to crystallise many copies of the molecule under investigation and effectively to use this ordered array as an amplifier of the scattering from a single molecule. To see atoms individually in less regular structures required the development of electron microscopy which has only recently reached the necessary resolving power. With this and other techniques, such as atomic force microscopy, the elaborate ordering of the atoms in both living and inorganic materials has begun to be revealed. In real materials we usually find several levels of organisation with different rules at each level. *Hierarchy* is the characteristic, in particular, of living systems. The present period has seen a rising appreciation of the ways in which structure and information are intimately connected. This was epitomised in the most important discovery of the twentieth century, the double helix of DNA, where the material structure encodes information as a sequence of base pairs. The general principle underlying this encoding of genetic information in an 'aperiodic crystal' had already been foreseen by Erwin Schrödinger in his book *What is Life?* (Schrödinger 1944). At the same time, immense computational power has developed, enabling the geometry of very complex structures to be handled and to be presented as computer graphics.

1.5 Shapes and Structures

The English and Scottish traditions of science, far more than the Continental, have been based on model-making, on *visualisation* and on analogy with everyday mechanisms rather than on words, formulae and logic-chopping. J. C. Maxwell, William Thomson (Lord Kelvin), W. H. D'Arcy Thompson, W. L. Bragg and J. D. Bernal are some of the masters of this British tradition, while Descartes, Gödel, Euler, Heisenberg, Claude Bernard, epitomise the Continental Schools, the attitude of which was exemplified by Pierre Duhem (1861–1916).

In 1917, in the middle of World War I, D'Arcy Wentworth Thomson in the University of Dundee produced his magnum opus *On Growth and Form*, in which he applied simple mathematics and physics to the problems of the multifarious shapes encountered in the living world.

It is clear that, even though considerable knowledge was available at the time, D'Arcy Thompson (1860–1948), did not wish to take atoms and microstructure at the atomic level into account. He mentioned the atomicity of matter only once or twice and even in the second edition (1942) took little note of the atomic level. His work was concerned with illuminating the underlying mathematical principles governing the shapes of living organisms at a macroscopic scale, stopping short at the level of things visible with a simple microscope.

A transitional manifesto, *Structure in Nature is a Strategy for Design* by an architect, Peter Pearce (1978) has been influential, just before the computer period, in introducing the geometry of polyhedra and related regular three-dimensional structures to a wider world so that the dominance of cubic structures in our culture has been reduced. Robert Williams' *The Geometrical Foundations of Natural Structure* (1979) takes a similar approach, taking examples of intricate polyhedral structures from the geometry of complex arrangement of atoms in crystalline materials. We must mention too the vital role played by H. S. M. Coxeter in reviving a general interest in geometry and educating several generations throughout his long life and particularly with his textbook *An Introduction to Geometry* (1969). Grünbaum and Shephard wrote: 'mathematicians have long since regarded it as demeaning to work on problems related to elementary geometry in two or three dimensions, in spite of the fact that it is precisely this sort of mathematics which is of practical value.'

Our aim, then, in the chapters which follow, is to survey some of the important developments that have been taking place in recent decades in our understanding of the structure of complex materials, with the emphasis on the underlying geometrical principles.

Johannes Kepler (1571–1630) can be regarded, for many reasons, as the initiator of this approach. It was Kepler who produced for the first time a rigorous enumeration of the 'regular' tilings of the plane. He also considered tilings that include regular pentagonal tiles and so came close to the concept of aperiodic patterns more than 300 years before the Penrose tilings and the discovery of quasicrystals. He rediscovered the 13 semi-regular polyhedra and discovered the rhombic triacontahedron, which is now known to be an important key to the structure of icosahedral quasicrystals. It was Kepler who suggested that the hexagonal symmetry of snowflakes could arise from a close packing of identical subunits, and it was Kepler who guessed, correctly, that the densest possible packing of spheres is the arrangement now known as 'cubic close packing' (Kepler 1611). Robert Hooke, applying the microscope for the first time to everything within his reach, began to realise a science of the structure of matter. Figure 1.3 is taken from Hooke's *Micrographia* (Hooke 1665).

We close this chapter with a picture (Figure 1.4) of a portion of Kepler's configuration of nested regular polyhedra, which he believed could account for the radii of planetary orbits. His unusual mental flexibility allowed him to abandon it and go on to

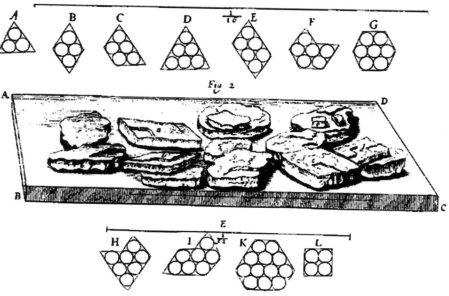

Figure 1.3 Crystalline structure from Robert Hooke's *Micrographia* (Hooke 1665).

discover the three famous fundamental laws of planetary motion. It may seem strange that we choose to pay tribute to Kepler's genius by thus drawing attention to a case where his remarkable intuition led him astray. However, the model epitomises the perennially fascinating attraction, for the human mind, of the five regular polyhedra. And the model now seems strangely prophetic – configurations of *nested regular and semi-regular polyhedra* have recently re-emerged in the striving of scientists to reveal Nature's structural principles, this time in a quite different context – as models of the clusters of atoms that occur in the building up of complex crystalline solids.

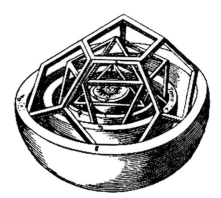

Figure 1.4 The portion of Kepler's configuration of nested regular polyhedra, representing the orbits of the inner planets. From Kepler's *Mysterium Cosmographicum* (1596).

2

2D Tilings

A *doubly periodic* pattern in the plane has a symmetry group containing two independent translations. Two translation vectors determine a parallelogram. A parallelogram whose sides have lengths and directions corresponding to two of the translations is a *primitive unit cell* of the pattern, provided there is no smaller parallelogram formed from translation vectors. Applying the translations to a single point generates a *lattice*. The whole pattern is produced by repeated application of the portion of the pattern within a primitive unit cell, by translating the cell. There are 17 discrete subgroups of the Euclidean group of the plane that contain two independent translations (see for instance Schattschneider (1978) for a simple introduction). They are the *wallpaper groups*, and patterns with these symmetry groups are 'wallpaper patterns'. In the standard nomenclature for these groups the symbol p denotes the *primitive* unit cell, and c denotes a longer rectangular unit cell whose vertices *and centre* are vertices of primitive parallelograms. The p or c is then followed by a list of generators: 2, 3, 4 and 6 denote rotations through $2\pi/2$, $2\pi/3$, $2\pi/4$ and $2\pi/6$, m denotes a reflection and g denotes a glide (the combined effect of a reflection in a line and a translation along the direction of the line and equal to half a lattice translation).

A *tiling*, or *tessellation*, of a space is a subdivision of the space into non-overlapping regions ('*tiles*'). The vertices and edges of a tiling constitute a *net*. Periodic nets in two or in three dimensions (2D or 3D) are of fundamental importance in the description of crystalline structure. In the simplest application vertices and edges corresponding to atoms and bonds, but highly complex structures can often be more readily understood and visualised by identifying an underlying net or framework in the structure – a topic we shall return to in a Chapter 8. O'Keeffe & Hyde (1980) have given an extensive and fascinating survey of 2D nets that occur in crystalline structures.

Doubly periodic patterns produced by tiling a plane have been exploited for their decorative possibilities by every civilisation, for thousands of years. The ingenuity of medieval Islamic craftsmen is particularly noteworthy, and often quite

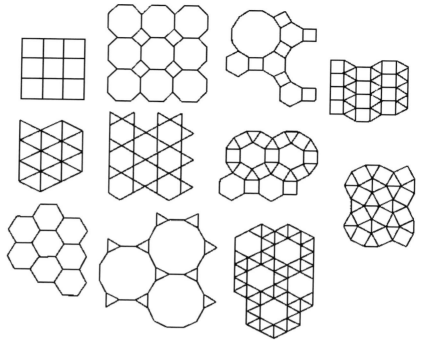

Figure 2.1 The eleven uninodal ways of tiling the plane with regular polygon tiles.

amazing (Bourgoin 1879; El-Said & Parman 1976; Critchlow & Nasr 1979; Chorbachi 1989).

2.1 Kepler's Tilings

The first known example of a mathematically rigorous approach to a tiling problem is probably Kepler's enumeration of all possible tilings of the plane by regular polygons, with the proviso that all vertices shall be identically surrounded (Kepler 1619).

Suppose that the generic vertex is surrounded by n_3 equilateral triangle, n_4 squares, and so on. The total angle around the vertex is 2π. This gives

$$\sum_{p=3}^{\infty} (p-2)n_p/p = 2$$

Eleven solutions lead to tilings of the whole plane. In an obvious notation, that lists the values of p encountered as the vertex is circumnavigated, we have:

$$3^6 \quad 4^4 \quad 6^3 \quad 3^4.6 \quad 3^3.4^2 \quad 3^2.4.3.4 \quad 3.4.6.4 \quad 3.6.3.6 \quad 3.12^2 \quad 4.6.12 \quad 4.8^2$$

Portions of these tiling patterns are illustrated in Figure 2.1. The tiling $3^4.6$ exists in two enantiomorphic versions. The tiling 3.6.3.6 is called the 'kagome' pattern

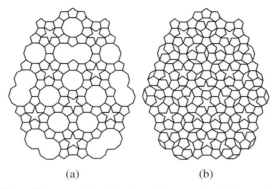

(a)   (b)

Figure 2.2 (a) A tiling illustrated by Kepler, (b) a patch of Penrose's aperiodic pentagon tiling.

Figure 2.3 A portion of Dürer's pentagon tiling.

because it resembles the traditional Japanese basket weave of that name. Kagome means 'dragon's eye' in Japanese.

Eight further solutions do not tile the whole plane; they leave gaps:

$$3.7.42 \quad 3.8.24 \quad 3.9.18 \quad 3.10.15 \quad 4.5.20 \quad 5^2.10 \quad 3^2.4.12 \quad 3.4.3.12$$

The solution $5^2.10$ leaves star-shaped, rhombic, and 'boat-shaped' gaps. Kepler actually illustrated a closely related pattern (Figure 2.2a). His remarks about it reveal that he was intuitively aware of the possibility of aperiodic tiling patterns, more than 350 years before Roger Penrose's discovery of a truly aperiodic tiling (Figure 2.2b) (Lück 2000). An interesting tiling of the plane by pentagons and 'lozenges' was proposed by Dürer (1523) (Figure 2.3). It is generated by an iterative procedure, beginning with a pentagon and at each stage placing pentagonal tiles on all the free edges of the growing patch.

2.2 Fundamental Regions

In E_2 the discrete symmetry groups are the 17 wallpaper groups and in S_2 they are the point groups, which are the symmetry groups of the cube, of the icosahedron, of the prisms and antiprisms, and the subgroups of these. Let G denote a discrete symmetry group of a space S and consider a tiling with the property that G acts transitively on the tiles (i.e. any tile can be mapped to any other by a transformation belonging to G) and that only the identity of G can map a tile to itself. A tile belonging to such a tiling is a *fundamental region* or *asymmetric unit* for the action of G on S. The importance of this concept comes from the fact that any configuration with symmetry G is completely determined by the portion of it that lies in a fundamental region. A fundamental region is a tile that, when decorated by an asymmetric motif and replicated by repeated action of the generating operations, produces the whole pattern. We illustrate this by a simple example. Figure 2.4 shows a binodal tiling (the dual of $3.4.3^2.4$) with symmetry group p4g and an asymmetric unit for the wallpaper group. The $45°$ right-angled triangle marked in dark grey is an asymmetric unit. (The lighter grey square is a unit cell.) Neighbouring units are related by reflection in the hypotenuse and quarter-turns about the vertex at the right-angle. Coxeter (1989) has listed a set of such generating transformations for each of the wallpaper groups. Their algebraic relations, which define the groups as abstract groups, are given in Coxeter & Moser (1957).

A new nomenclature for the 17 plane groups, that applies equally well to the symmetry groups of S_2 or H_2, is the *orbifold notation* (Conway 1992; Conway & Huson 2002; Hyde & Ramsden 2000a). Any of these 2D symmetry groups can be denoted by a list of generators (rotational symmetries, reflections and glides). In the orbifold notation a $2\pi/n$ rotation about a point lying on a reflection line is denoted by the symbol n to the right of an asterisk. Symbols to the left of the asterisk denote rotations not lying on lines of reflection. The group p4g (Figure 2.4), for example, is 4*2 in the orbifold notation. An *orbifold* is actually a topological object obtained by identifying edges of the asymmetric unit that are related by a rotation indicated by a symbol on the left of the asterisk. The concept of orbifolds has been extended to 3D and leads to a deeper understanding of the topological properties of the space groups (Delgado-Friedrichs & Huson 1997; Johnson *et al.* 2002).

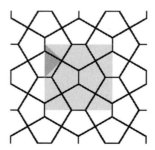

Figure 2.4 Unit cell and asymmetric unit for a tiling with symmetry p4g.

2.3 Topology of Plane Tilings

Consider a doubly periodic tiling of E_2 for which the net of vertices and edges is *c-connected* – that is, every vertex belongs to c edges. Suppose, further, that the pattern consists of F_3 triangles, F_4 quadrilaterals, … F_n n-gons, per unit cell, and that the net of edges and vertices is *c-connected* – that is, every vertex belongs to c edges. Then:

$$\sum_{n=3}^{\infty} \{2c - n(c - 2)\} F_n = 0$$

It is easy (but instructive) to verify this formula, for the Kepler tilings. Of course, the formula is concerned only with topology, not with metrical properties, so if we do not insist that the tiles be *regular* polygons further interesting possibilities arise. Figure 2.5 illustrates a plane tiling of pentagons and heptagons ($c = 3$, $F_5 - F_7 = 0$).

Defects in the form of pentagonal and heptagonal rings can occur in graphite layers (whose defect-free structure takes the form of the regular tiling 6^3). A 3-connected tiling of E_2 consisting of pentagons, hexagons and heptagons must satisfy $F_5 = F_7$. The introduction of a pentagon and a heptagon into the 2D honeycomb pattern 6^3 produces a dislocation with an associated Burgers vector whose magnitude and orientation depends on the relative positions of the pentagon and heptagon. For a defect in a honeycomb pattern consisting of two such pentagon–heptagon pairs the effect can be cancelled, so that no dislocation is produced. The simplest case is the Stone–Wales defect (Stone & Wales 1986) (Figure 2.6).

A carbon nanotube is basically a graphite sheet wrapped into cylindrical shape. It can occur bent or twisted, depending on the arrangement of the pentagonal, heptagonal and octagonal 'defects'. Terrones and Terrones (2003) have given the name 'Haeckelites' to exotic carbon nanotubes in which heptagonal and pentagonal rings predominate – the name was inspired by the beautiful drawings of radiolaria published by the German zoologist, Ernst Haeckel.

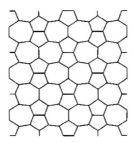

Figure 2.5 One of several ways of tiling a plane by equal numbers of pentagons and heptagons.

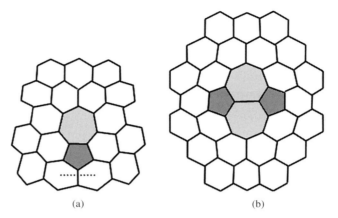

(a)   (b)

Figure 2.6 (a) A pentagon and a heptagon in a hexagonal tiling pattern. The Burgers vector associated with the dislocation is indicated by the dotted line near the base of the figure, (b) a Stone-Wales defect in a hexagonal tiling pattern. Notice that it can be produced simply by rotating one edge of the pattern.

2.4 Coloured Symmetry

A *monohedral* tiling is one in which all the tiles are congruent. Figure 2.7 shows a monohedral tiling pattern with black and white tiles. Black and white are interchanged by a fourfold rotation. The full symmetry group of the tiling is p4; the subgroup that does not interchange black and white is p2. As was first shown by H. J. Woods (1936) and Crowe (1986), there are 46 types of doubly periodic two-colour 'counterchange' patterns in E_2. Schattschneider (1986) has given an interesting introduction to methods of construction. They can be classified by *pairs G–H* of wallpaper groups, where *H* is a subgroup of *G* of index 2. (An interesting 3D analogue of this arises in the classification of triply periodic 'balance' surfaces.)

More generally, one can consider *n*-colour patterns in E_2 ($n = 2, 3, 4$ or 6) whose symmetry group *G* allows permutation of the colours and its subgroup of

Figure 2.7 A counterchange pattern of type p4/p2.

index *n* does not. This interesting extension of the concept of symmetry, in 2D and 3D, is explored in the book *Coloured Symmetry* (Shubnikov & Belov 1964). A simple introduction to the group theory involved has been given by Coxeter (1987). The concept can be extended to quasiperiodic patterns (Lifshitz 1996; 1998; Scheffer & Lück 1999).

2.5 Truchet Tilings

The idea of encoding a structure as a symbol string, from which the structure can be reconstituted, is of fundamental and far-reaching importance. The symbol string can be thought of as a 'gene' for the structure. Sabastian Truchet's approach is a very early example of this way of thinking about structures.

A rich variety of amazingly complex patterns can arise from applying very simple rules to very simple building units. A striking example of this is provided by the tiling patterns investigated by Sebastian Truchet (1704). They are constructed from a single square tile decorated by distinguishing the two halves above and below a diagonal (say, for definiteness, black and white). When a regular square tiling is built from tiles of this kind, there is a choice of four different orientations for each tile. Figure 2.8a shows an example. Labelling the four orientations A, B, C, D, the unit cell of this pattern can be specified by the notation:

AABDCCDB

DBAABDCC

CCDBAABD

BDCCDBAA

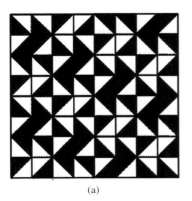 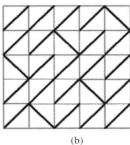 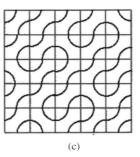

    (a)   (b)   (c)

Figure 2.8 (a) A tiling produced by the method of Sebastian Truchet, (b) a variant of a Truchet pattern produced by eliminating the distinction between black and white regions, (c) an example of a tiling by Cyril Smith's variant.

(The figure shows two rectangular unit cells. Observe, incidentally, that this particular example is a 'counterchange' pattern.) Truchet illustrated a wide variety of fascinating patterns. His work has been translated by Cyril Stanley Smith (1987). Examples of the patterns can be found on the Internet. All of Truchet's examples were *periodic* patterns, but this is of course not a necessary restriction – random patterns and quasi-periodic patterns can be just as readily generated from the ABCD notation.

In the pattern shown in Figure 2.8b the Truchet tile has been simplified still further; the square tile is decorated by a single diagonal. There are now only two possible orientations, which may be denoted by L and R. The pattern shown has a 5×5 unit cell and is obtained by shifting the row sequence … RRRRL … two tiles to the left, with respect to the previous row.

Smith introduced a variant of the tile in which the decoration is a pair of circular arcs. Again, there are just two orientations. The pattern generated by the periodic code of the previous example now looks like Figure 2.8c.

2.6 Aperiodic Tilings

Penrose's important discovery of aperiodic tiling patterns employing only two tile shapes was first published, not by Penrose, but by Martin Gardner (Gardner 1977; Penrose 1978). See Chapter 10 of Grünbaum & Shephard (1987) and Gardner's 1977 article for detailed properties, or Lord (1991) for a brief overview of the equivalencies among the several variants of Penrose tilings. The three most commonly encountered versions of the Penrose aperiodic patterns are the 'rhomb' patterns, the 'kite and dart' patterns and the 'pentagon patterns'. These are illustrated in Plate I, along with superimpositions in pairs to exhibit their essential equivalence – each version can be converted to each of the others by a simple rule.

A portion of a Penrose rhomb pattern is shown in Figure 2.9a, with the edges of the two kinds of tile marked by two kinds of arrowheads. A *matching rule* that ensures that no periodic tiling of the plane is possible is provided by marking the tile edges by two kinds of arrowheads as shown in the figure. For the markings in Figure 2.9b, discovered by Robert Ammann (a highly creative and original thinker in the field of 2D and 3D aperiodic tilings, who never published his results) the matching rule requires that, when two tiles are placed edge-to-edge, the lines decorating the tiles have to continue undeviated across the common edge. This produces a pattern of five sets of parallel lines ('Ammann bars').

Aperiodic *patterns* in general can be generated from the *decoration* of an aperiodic set of tiles – in the same way that periodic patterns are produced from the decoration a fundamental region of a wallpaper group. A striking example is the

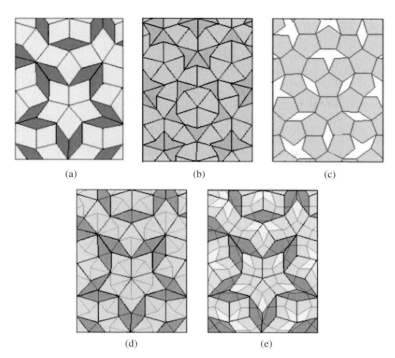

Plate I Penrose tiling patterns. (a) Rhomb pattern; (b) kite and dart; (c) pentagon pattern. (d) Superimposition of (a) and (b) – observe that all the 'fat rhombs' are decorated identically by the kite and dart pattern, as are all the thin rhombs. (e) The superposition of (a) and (c), demonstrating the equivalence of the rhomb and the pentagon patterns.

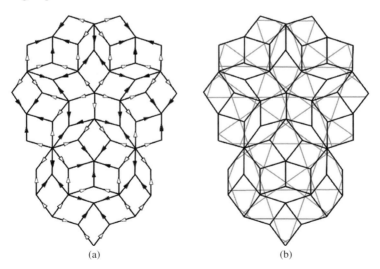

Figure 2.9 (a) The standard marking of the edges of the two Penrose rhombs, that establishes the matching rule; (b) a portion of a Penrose rhomb pattern marked by Ammann bars.

Figure 2.10 An aperiodic weaving pattern with five thread directions.

'weaving' pattern of Figure 2.10. Observe that all the fat rhombs are identically decorated, as are all the thin rhombs. The net result is a generalisation of the 'under and over' pattern of fibres in weaving. Usually there are two sets of fibres (warp and weft) in two directions. In the traditional Japanese *kagome* basket weave (the 3.6.3.6 pattern of Figure 2.1) there are three. Here, there are five (Mackay 1988). Plate II illustrates an aperiodic weave with five thread directions in which the threads correspond to the Ammann bars of a Penrose pattern.

Petra Gummelt has described aperiodic pairs of tiles with fractal boundaries (Gummelt 1995). The two-tile patterns are closely related to Penrose's pentagon patterns (illustrated in Figure 2.2b, with the markings that specify the matching rules that force aperiodicity omitted) were discovered somewhat earlier (Penrose 1974). The rhomb patterns can be generated either by marking the edges of the tiles and imposing matching rules, by projecting a slice of a 5D hypercubic lattice onto E_2 (Kramer & Neri 1984; Conway & Knowles 1986), or by inflation rules (Grünbaum & Shephard 1987). The projection method is essentially equivalent to de Bruijn's 'pentagrid' method (De Bruijn 1981). The Penrose patterns are characterised by a long-range orientational order with fivefold symmetry – a symmetry incompatible with periodicity. Much of the early theoretical work on quasicrystals was inspired by the Penrose tiling patterns and their 3D analogues (Steinhardt & Ostlund 1987).

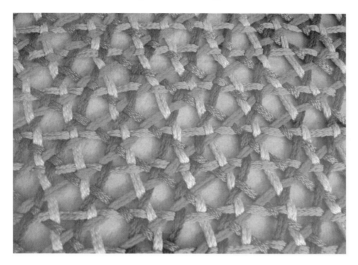

Plate II Five-stranded weaving pattern based on the Penrose rhomb tilings. Detail from a coffee table designed and made by Robert Mackay.

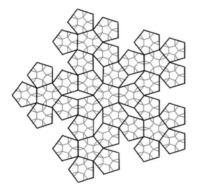

Figure 2.11 Central patch of a hierarchical pattern of regular pentagons.

The set of tiles that constitute a Penrose tiling pattern is an example of an *aperiodic set of tiles*, by which is meant a set of tiles that can tile the plane but from which no periodic tiling pattern can be built. Less restrictive is simply the concept of a tiling that is not periodic. Some examples of this are of special interest. A hierarchical pattern of pentagonal tiles is illustrated in Figure 2.11. Starting from a single pentagon, it can be produced by an iterative procedure; at each iteration every pentagonal tile is converted to six smaller pentagonal tiles. The gaps between pentagons, as well as the pentagons themselves, can be tiled by just two tiles (an isosceles triangle with base angle $\pi/5$ and a base equal to $\tau = (1 + \sqrt{5})/2$ times the pentagon edge, and an isosceles triangle with a base angle $2\pi/5$ and base τ^{-1}), thus producing a non-periodic tiling of the plane (Figure 2.12) (Mackay 1976). A curious monohedral tilings of the

Figure 2.12 A larger portion of the hierarchical pentagon pattern, with the gaps between pentagons tiled by pentagons and two kinds of triangular tile.

Figure 2.13 A Voderberg spiral monohedral tiling.

plane, in which the tiles are arranged spirally (Figure 2.13), was discovered by Voderberg (1937). Further examples of spiral tiling patterns are to be found in Fejes Tóth (1964) and Grünbaum & Shephard (1981; 1987).

2.7 Gummelt's Decagon

If the 'tiles' are allowed to overlap, rather than abutting edge to edge, the result is a *covering* of the space. Gummelt (1995; 1996; Gummelt & Bandt 2000; Jeong & Steinhardt 1997) discovered that, in the case of covering of E_2, a *single tile* with determinate internal structure suffices to force aperiodicity. Gummelt's tile is a

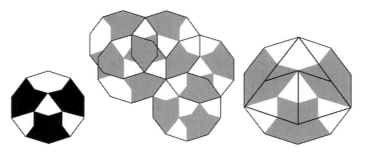

Figure 2.14 (a) Gummelt's decagon, (b) a portion of an aperiodic pattern of overlapping decagons, (c) a correspondence between Gummelt's decagon patterns and Penrose's kite and dart patterns.

regular decagon, decorated by black and white regions as in Figure 2.14a. the simple matching rule 'black on black, white on white', together with the requirement that every decagon edge is covered by the interior of another decagon, suffices to ensure an aperiodic pattern (Figure 2.14b). The resultant patterns, as Gummelt showed, are equivalent to the Penrose tiling patterns. The simplest demonstration of the equivalence is indicated in Figure 2.14c, where two kites and a dart have been superimposed on a Gummelt decagon (Lord *et al.* 2000; 2001). However, this does not necessarily imply the equivalence of patterns based on decorating the tiles of a Penrose pattern and those produced by decorating the Gummelt decagons. Some interesting aspects of this problem have been studied (Lord & Ranganathan 2001a, b; Jeong 2003). The method of Janot & Patera (1998) for generating aperiodic point sets by overlapping decagons is closely related to the Gummelt covering. A remarkable model of the decagonal Al–Ni–Co quasicrystal was proposed by Steinhardt *et al.* (1998), in which the atoms decorate the upper and lower surfaces of a decagonal prism. The structure is generated by Gummelt's matching rules. Several other decagonal phases can be constructed in a similar way (Lord & Ranganathan 2001a, b).

2.8 The Divine Proportion and the Fibonacci Sequence

The *golden number* tau,

$$\tau = (1 + \sqrt{5})/2 = 1.618033989 \ldots$$

(Pacioli 1509; Dunlap 1998; Herz-Fischler 1998; Livio 2003) (sometimes also called phi, ϕ) is fundamental to the geometry of fivefold symmetry and tenfold symmetry (Hargittai 1992), since it is the length of a diagonal of a regular pentagon of unit edge length and also the radius of the circumscribing circle of a regular decagon of unit edge length. Since τ is, for this reason, intimately involved in the properties of the Penrose tilings, this seems a convenient place to enter into a short digression on its properties. We shall meet it again in later chapters.

The ratio $1 : \tau$ occurs in Euclid as the solution to the problem of dividing a line into two parts so that the ratio of the small part to the large part is equal to the ratio of the large part to the whole line. This implies that τ is the positive root of $x^2 - x - 1 = 0$. The diagonals of a regular pentagon divide each other in this ratio. The ratio has a place in the history of art and architecture, being associated with aesthetically pleasing proportions (Ghyka 1946; Jeanneret 1954; Huntley 1970; Elam 2001). Luca Pacioli's book *De Divina Proportione* (Pacioli 1509) dealing with the properties of this number was illustrated by Leonardo da Vinci.

The Indian method of notation for numbers, adopted by the Islamic world and now known as 'Arabic numerals', was introduced into Europe by Leonardo of Pisa – nicknamed 'Fibonacci' – in his book *Liber Abaci* ('book of the abacus') (Pisano 1202). The new notation eventually overthrew the clumsy Roman notation and made the abacus obsolete. In *Liber Abaci* Fibonacci considered the sequence:

$$1\ 1\ 2\ 3\ 5\ 8\ 13\ 21\ 34 \ldots$$

(now known as the *Fibonacci numbers*), where each number is the sum of the two that precede it:

$$f_1 = f_2 = 1, f_{n+1} = f_n + f_{n-1}$$

He explained the sequence in terms of the multiplication of a population of rabbits. Start with a pair of baby rabbits, denoted by B. At the end of a month, they have become adults: $B \to A$. At the end of the next month the adult pair produces a new pair of baby rabbits: $A \to AB$. The rabbits are supposed to be immortal, so that we get the sequence:

$$B \quad A \quad AB \quad ABA \quad ABAAB \quad ABAABABA \quad ABAABABAABAAB \ldots$$

Observe that the population grows month by month in accordance with the Fibonacci number sequence. These letter sequences have been much investigated in connection with quasicrystal theory – in the limit the infinite symbol string is *aperiodic* – a 'one-dimensional (1D) quasicrystal'. Note its *hierarchical* structure – each of the sequences in this sequence of sequences is obtained by joining the two preceding sequences end to end.

The Penrose rhombic tiling pattern is in fact the 2D generalisation of this 1D structure. The Ammann bars associated with a Penrose pattern constitute five sets of parallel lines; within each set, the spacing conforms to the sequence discussed above, with A and B corresponding to a distance τ and a distance 1, respectively.

The Fibonacci number sequence has many surprising properties (Dunlap 1997). Of particular relevance to the geometry of quasicrystals and their crystalline approximants is the fact that the ratios of pairs of successive terms converge, in the limit, to the golden number:

$$\frac{1}{1}, \frac{2}{1}, \frac{3}{2}, \frac{5}{3}, \frac{8}{5}, \frac{13}{8}, \frac{21}{13}, \dots . \tau$$

We thus get a sequence of rational approximants to the irrational number τ. This result (due to Kepler) is a consequence of the curious form of the golden number, when expressed as a continued fraction:

$$t = 1 + \cfrac{1}{1 + \cfrac{1}{1 + \cfrac{1}{1 + \dots}}}$$

This, in turn, is readily obtained from the quadratic equation, rewritten in the form:

$$\tau = 1 + \frac{1}{\tau}$$

2.9 Random Tilings

Random tilings of E_2 or E_3 occur naturally in various situations: in biological cell structures, in geological formations, in grain structures of metals and ceramics, and in foam. There are striking statistical laws that reveal an underlying similarity of these widely different phenomena (Kikuchi 1956; Weaire & Rivier 1984).

2.10 Spherical Tilings

The few basic ideas on tilings in E_2 presented in the foregoing sections are sufficient for our purposes. A very thorough presentation of the intricate field of tilings in E_2 is Grünbaum & Shephard's *Tilings and Patterns* (1987). We now consider tilings in S_2 (i.e. the partitioning of the surface of a sphere into polygonal regions).

Kepler proved that the only finite polyhedra with regular polygon faces and all vertices identically surrounded are the five *regular* polyhedra, or 'Platonic solids'

$$\{3, 3\} \ \{3, 4\} \ \{4, 3\} \ (3, 5) \ \{5, 3\} \ (\text{equivalently, } 3^3 \ 3^4 \ 4^3 \ 3^5 \ 5^3)$$

(Figure 2.15) the 13 Archimedean solids:

$$3.6^2 \quad 3.8^2 \quad 3.4.3.4 \quad 4.6^2 \quad 4.6.8 \quad 3.4^3 \quad 4.3^4 \quad 5.6^2 \quad 3.12^2$$
$$3.5.3.5 \quad 3.4.5.4 \quad 4.6.10 \quad 5.3^4$$

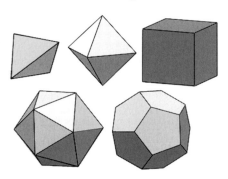

Figure 2.15   The five Platonic solids or regular polyhedra: tetrahedron, octahedron, cube, icosahedron and dodecahedron.

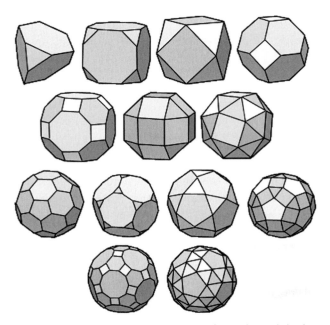

Figure 2.16   The 13 Archimedean solids or semi-regular polyhedra.

(Figure 2.16) (the snub cube 4.3^4 and the snub dodecahedron 5.3^4 each exist in two enantiomorphic versions), and the prisms and antiprisms $4^2.p$ and $3^3.p$. For recent publications dealing with the Platonic and Archimedean solids, see Cromwell (1999) and Sutton (2002). Kepler's work on tilings and polyhedra is discussed by Cromwell (1995; 1999). The Archimedean solids, the prisms and the antiprisms together constitute the class of *semi-regular* polyhedra. Models of these figures can be constructed from a flat sheet by folding (Cundy & Rollet 1961; Williams 1979; etc.) – an idea that first occurs in Albrecht Dürer's *Unterweysung der Messung*

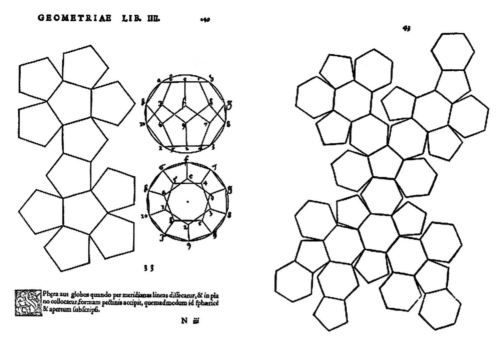

Figure 2.17 Two pages from Dürer's *Unterweysung der Messung* showing the planar figures that can be folded to form (a) a dodecahedron and (b) a truncated icosahedron.

Figure 2.18 Spherical tilings. The triangular tiles are the fundamental regions for patterns with (a) tetrahedral, (b) cubic and (c) icosahedral symmetry.

(Dürer 1523), where flat nets for the Platonic solids are illustrated (Figure 2.17). Each of these 3D figures has all vertices on a circumscribing sphere; it is often convenient to consider the equivalent tiling of the surface of a sphere S_2 (Figure 2.18).

The concept of semi-regular polyhedra can be generalised by allowing regular 'star' polygons as faces, and allowing faces to intersect each other, as in the regular polyhedra {5/2, 3} and {5/2, 5} (Kepler 1619) and their duals {3, 5/2} and {5, 5/2} (Poinsot 1810). A survey of generalised semi-regular polyhedra has been provided

Figure 2.19   Successive truncations of a cube.

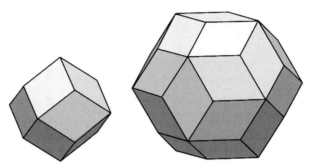

Figure 2.20   The rhombic dodecahedron and the rhombic triacontahedron.

by Coxeter *et al.* (1953), and the list was shown by Skilling (1975) to be complete. Wenninger (1971; 1983) has produced models and described methods of construction for these objects and their duals.

Two interesting ways in which tilings or polyhedra can be related to each other is through the process of *truncation*, or by the *duality* relation. Figure 2.19 illustrates the sequence of polyhedra obtained by successive truncation of a cube: 4^3 (cube) – 3.8^2 (truncated cube) – 3.4.3.4 (cuboctahedron) – 4.6^2 (truncated octahedron) – 3^4 (octahedron).

The Archimedean polyhedra that are of particular importance because of their prominent role in material structures are: the truncated tetrahedron or Friauf polyhedron 3.6^2, the cuboctahedron 4.6.4.6, the truncated octahedron 4.6^2, and, since the discovery of C_{60}, the icosidodecahedron 5.6^2.

Two 2D tilings or polyhedra are *dual* to each other if each face of one contains just one vertex of the other and each pair of vertices connected by an edge in one figure correspond to a pair of faces sharing a common edge, in the other. Thus, for example, the cube 4^3 and octahedron 3^4 are duals, as are the icosahedron 3^5 and the dodecahedron 5^3; the tetrahedron 3^3 is self-dual. The duals of the Archimedean solids are sometimes referred to as the Catalan solids (Catalan 1865). Their faces are congruent but not regular. Two important cases are the *rhombic dodecahedron* (dual of the cuboctahedron 3.4.3.4) and the *rhombic triacontahedron* (dual of the icosidodecahedron 3.5.3.5). These are illustrated in Figure 2.20.

Kepler's 11 tilings of E_2 and 13 tilings of S_3 are *vertex transitive*, or *uninodal*, meaning that any vertex can be taken to any other by a symmetry of the tiling – the vertices are all *equivalent* under the action of the symmetry group. Their duals are, correspondingly, *face transitive*. The vertices and edges of a tiling of a 2D space constitute a 2D *net*. A net is *uninodal, binodal, trinodal*, etc. according as the number of inequivalent vertices is one, two, three, etc. Note that these concepts are topological concepts; the symmetry group of a topological net is the group of permutations of vertices that also permutes the edges.

2.11 Topology of 2D Tilings

The numbers of vertices, edges and faces of a tiling of a sphere satisfy the well-known Euler relation $V - E + F = 2$. For the more general case of a tiling on a closed surface of genus g:

$$V - E + F = 2 - 2g = \chi$$

χ is the Euler characteristic. For a tiling all of whose vertices are *c*-connected, consisting of F_3 triangles, F_4 quadrilaterals,... F_n *n*-gons, we have the further relations:

$$F = \Sigma F_n, \qquad 2E = \Sigma n F_n = cV$$

(i.e. every edge belongs to two vertices and two tiles, every vertex belongs to *c* edges and every *n*-gonal tile has *n* vertices). It follows that, for a *c*-connected tiling on a surface of genus *g*:

$$\Sigma \{ 2c - n(c - 2) \} F_n = 2c\chi$$

The formula given in Section 2.4 is a particular case of this more general result: for a doubly periodic tiling pattern on the Euclidean plane, one can consider the unit cell to be topologically a torus (by identifying pairs of opposite edges). Its Euler characteristic is therefore zero.

Tilings of a topological sphere for which all vertices have the same connectivity must satisfy:

$(c = 3)$ $\qquad\qquad 3F_3 + 2F_4 + F_5 - F_7 - 2F_8 - \cdots = 12$

$(c = 4)$ $\qquad\qquad\qquad F_3 - F_5 - 3F_7 - \cdots = 8$

$(c = 5)$ $\qquad\qquad F_3 - 2F_4 - 5F_5 - 8F_6 - \cdots = 20$

In particular, the formula admits a 3-connected tiling of a sphere by 12 pentagons (corresponding to the pentagonal dodecahedron). More interestingly, the formula allows tiling of a sphere by 12 pentagons and an *unspecified* number of hexagons.

2.12 Tilings and Curvature

The topics discussed in the previous two sections indicate an intimate relationship between the topological properties of a tiling on a curved surface and the curvature properties of the surface. Why this should be so is clarified by the Gauss–Bonnet theorem. A surface has two *principal curvatures* at each point; these are the maximum and minimum curvatures of the curves of intersection of the surface with a plane through its normal. The *Gaussian curvature K* is their product. Negative Gaussian curvature at a point implies that the surface is 'saddle-shaped' in the neighbourhood of the point.

Consider an *n*-gon on a curved surface, whose edges are geodesics. Let Θ be the sum its interior angles. The Gauss-Bonnet theorem is essentially the equality:

$$\int_A KdS = \Theta - (n - 2)\pi$$

(The integration is over the area of surface contained in the polygon.) For a tiling on a closed surface we can sum over all the tiles. Since $\Sigma\Theta = 2\pi V$ and $\Sigma(n - 2)F_n = 2E - 2F$, we have the Gauss-Bonnet theorem relating the integrated Gaussian curvature to the Euler characteristic:

$$\int_S KdS = 2\pi(V - E + F) = 2\pi\chi$$

Let us now suppose the geodesic polygon is a tile belonging to a 3-connected tiling of the surface, and make the simplifying assumption that the edges at each vertex meet at approximately $2\pi/3$. Then:

$$(3/\pi)\int_A KdS = 6 - n$$

Thus, tiles with $n < 6$ contribute to regions of positive curvature (12 pentagons being just enough to curve the surface into a topological sphere – as in the fullerenes), while those with $n > 6$ contribute to regions of negative curvature. The net in Figure 2.21 illustrates this. It has hexagons and octagons, and lies on a surface of negative Gaussian curvature (actually a pair of unit cells of the triply periodic P-surface which will be dealt with in more detail in Chapter 9). The hypothetical carbon materials consisting of negatively curved graphite sheets to form such 3D triply periodic structures are called *Schwarzites* (Mackay & Terrones 1991; 1993). Further extensions of the principle whereby graphite sheets can be given curvature by introducing pentagons and heptagons have been explored, and interesting geometrical configurations, both observed and hypothetical, have been described (Terrones & Terrones 1996; 2003; Terrones *et al.* 2004).

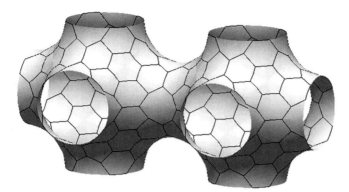

Figure 2.21 An example of a hypothetical material known as a 'Schwarzite': a continuous triply periodic surface formed from a graphite sheet. The octagonal rings among the hexagons permit the negative Gaussian curvature.

2.13 Fullerenes

The 60 vertices of the Archimedean polyhedron 5.6^2 (the truncated icosahedron) correspond to the positions of carbon atoms in the fullerene C_{60} (Kroto *et al.* 1985). Larger defect-free fullerenes also correspond to tilings of spheres by hexagons and pentagons (Smalley & Curl 1991; Chung & Sternberg 1993). According to the formula for 3-connected tilings the number of pentagons is necessarily 12. The number of hexagons may vary.

Given any 3-connected spherical net (or, for that matter, any 3-connected net on any surface), there are two iterative procedures that will introduce more hexagonal tiles, as illustrated in Figure 2.22. Each iteration of the first procedure changes the numbers of vertices, edges and faces according to $V \to 4V$, $E \to 4E$, $F_6 \to F_6 + E$ (other F_n unchanged); each iteration of the second procedure gives $V \to 2E$, $E \to 3E$, $F_6 \to F_6 + V$ (other F_n unchanged). These two operations commute. Starting from the dodecahedron and applying these two procedures we can deduce the possible existence of the fullerenes with exact icosahedral symmetry listed below:

20, 30, 12	80, 120, 42	320, 480, 162	1280, 1920, 642 ...
60, 90, 32	240, 360, 122	960, 1440, 482	...
180, 270, 92	720, 1080, 362	...	
540, 810, 272	2160, 3240, 1082	...	
1620, 2430, 812	...		
...			

Each entry gives V, E, F. A step to the right corresponds to the first type of iteration, a step downwards corresponds to the second type of iteration. The three cases $V = 60$, 80 and 180 are shown in Figure 2.23.

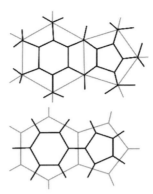

Figure 2.22   Two iterative procedures applied to a 3-connected tiling.

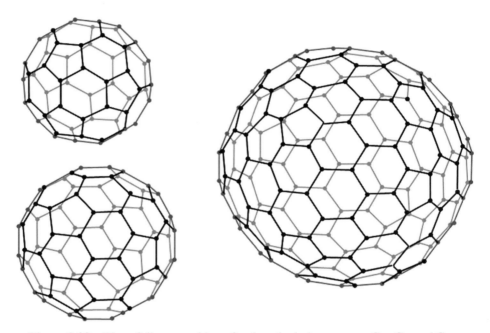

Figure 2.23   Three fullerenes with perfect icosahedral symmetry: C_{60}, C_{80} and C_{180}.

Applying the first iteration m times and the second iteration n times, starting from the pentagonal dodecahedron, gives a polyhedron with hexagonal and 12 pentagonal faces, with:

$$V = 20N, E = 30N, F = 2 + 10N$$

where $N = 4^m 3^n$. This accounts for all fullerenes that have exact icosahedral symmetry. Of course, less symmetrical fullerene molecules exist, the simplest of which is C_{70}.

Defects in the form of pentagonal and heptagonal rings can occur in graphite layers. The overall flatness of the layer is unaffected so long as $F_5 = F_7$. Similarly, defects in fullerenes in the form of heptagonal rings can occur – so long as they are compensated by the same number of additional pentagons, the closed shape (topological sphere) is unaffected. Indeed, the larger 'perfect' fullerenes with 12 pentagons tend to be less spherical; lower energy configurations that are more nearly spherical contain heptagonal (and additional pentagonal) rings (Terrones & Terrones 2003).

Carbon in the form of graphite is not the only crystalline material that consists of layers with a hexagonal pattern, in which the layers are loosely bound to each other. It might be expected that other materials with this kind of layered structure might produce fullerene-like molecules, wherein a layer is curved into a sphere by the presence of defects. Certain metallic disulphides that consist of layers of hexagonal rings have been observed to form such nanoparticles (Tenne *et al.* 1992; Margulis *et al.* 1993; Tenne *et al.* 1998; Terrones & Terrones 2003).

3

3D Tilings

In this chapter, and elsewhere, we shall employ the Herman–Mauguin (H-M) nomenclature for the space group symmetries of triply periodic structures. The definitive source for complete information on the space groups, *The International Tables for Crystallography* (Hahn 1995) is rather daunting for the beginner. An excellent explanation of the groups and the H-M symbols can be found in Phillips (1946, 1956). A very detailed explanation cannot be provided here, for reasons of space limitations, but a brief indication of the meaning of the H-M notation is in order, to render the H-M names of the space groups less mysterious to the uninitiated. The following rather lengthy section could be omitted on a first reading without seriously affecting comprehensibility of the rest of the chapter.

3.1 Lattices and Space Groups

A *lattice* in Euclidean 3-space is an array of points whose position vectors are of the form $n_1\mathbf{t}_1 + n_2\mathbf{t}_2 + n_3\mathbf{t}_3$ where n_1, n_2 and n_3 are integers and $\mathbf{t}_1$, $\mathbf{t}_2$ and $\mathbf{t}_3$ are three independent vectors. The vectors $n_1\mathbf{t}_1 + n_2\mathbf{t}_2 + n_3\mathbf{t}_3$ specify the translational symmetries of the lattice. There are 14 kinds of lattices in three dimensions (3D); they are shown in Figure 3.1. Each type is known as a *Bravais lattice*.

Three lattice vectors **a**, **b** and **c** can be selected to provide a reference system. They need not be the three edges ($\mathbf{t}_1$, $\mathbf{t}_2$ and $\mathbf{t}_3$) of a *primitive* cell. The reference

Cubic	$a = b = c$	$\mathbf{a.b} = \mathbf{a.c} = \mathbf{b.c} = 0$
Hexagonal	$a = b$	$\mathbf{a.b} = (-1/2)a^2$, $\mathbf{a.c} = \mathbf{b.c} = 0$
Tetragonal	$a = b$	$\mathbf{a.b} = \mathbf{a.c} = \mathbf{b.c} = 0$
Orthorhombic		$\mathbf{a.b} = \mathbf{a.c} = \mathbf{b.c} = 0$
Monoclinic		$\mathbf{a.b} = \mathbf{b.c} = 0$
Triclinic	*no special restrictions*	

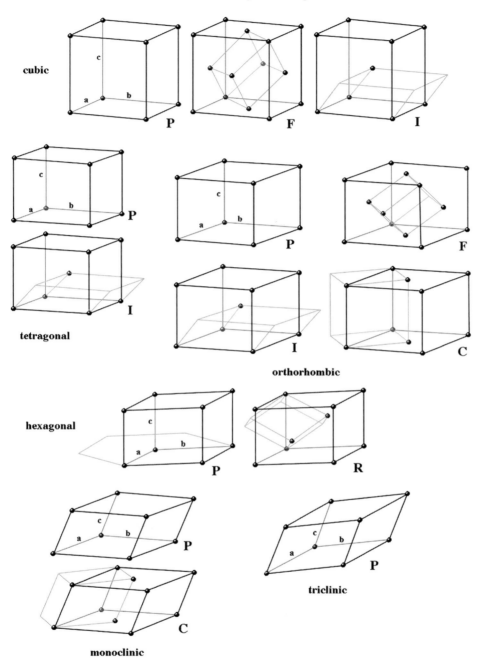

Figure 3.1 The 14 Bravais lattices. The unit cells outlined in black are the ones conventionally employed. The letters P, F, I and C mean primitive, face-centred, body-centred (innenzentriert) and base-centred. R indicates the existence of a rhombohedral unit cell. The smaller cell outlined in light grey are primitive cells.

systems conventionally used are those indicated in Figure 3.1. Observe the special relationships among these three vectors, in the various cases:

Any triply periodic structure (crystalline structure) has, by definition, three independent translational symmetries. Hence any triply periodic structure is associated with one of the 14 Bravais lattices. It may also have *fixed point symmetries*: rotations about axes through the fixed point, reflections in planes through the fixed point, and various combinations of these (inversion, rotary-inversion and rotary-reflection). There are just 32 groups of fixed point symmetries that are compatible with triple periodicity. These are the *point groups*. They can be classified into seven *crystal systems*, according to the number and arrangement of their rotational axes. A *triclinic* point group has no rotational symmetries, a *monoclinic* point group has just *one* axis of twofold rotation (*diad*), an *orthorhombic* point group has three mutually perpendicular diads. A tetragonal point group has one fourfold axis (*tetrad*) and a hexagonal point group has one threefold axis (*triad*) or a sixfold axis. The cubic groups are characterised by having four independent triads.

A point group is named by a string of up to three symbols. The symbols represent axes and planes through the fixed point. A number n denotes an axis of rotation through $2\pi/n$, where $n = 2, 3, 4$ or 6 (1 indicates the absence of any rotation axis and, as is well known, 5 is forbidden because it is incompatible with a lattice). $\bar{1}$ indicates that the fixed point is an inversion centre, and $\bar{2}, \bar{3}, \bar{4}$ and $\bar{6}$ refer to rotary-inversion axes. A reflection plane is denoted by m. A number and an m separated by a slash refers to an axis and a mirror plane perpendicular to it (e.g. 3/m, $\bar{4}$/m); in the following explanation this combination is regarded as a *single symbol*. The *positions* of the symbols in the name of a point group indicates the *relative orientations* of the axes and normals to the planes. The adopted convention can be summarised in the table below:

Reference lattice	System	First position	Second position	Third position
Cubic	Cubic	**c**	**a + b + c**	**a − b**
Hexagonal	Hexagonal or trigonal	**c**	**a**	**a − b**
Tetragonal	Tetragonal	**c**	**a**	**a − b**
Orthorhombic	Orthorhombic	**a**	**b**	**c**
Monoclinic	Monoclinic	**b**		

The crystal system to which a particular point group belongs can be seen from the name given to it according to this convention. If the *second* symbol in the name is 3 (or $\bar{3}$) the class is *cubic*. Otherwise, it is *triclinic, trigonal, tetragonal* or *hexagonal* according as the first symbol is 1, 3, 4 or 6 (or $\bar{1}, \bar{3}, \bar{4}$ or $\bar{6}$). If none of these is the case and there is *only one* symbol in the name (2m or 2/m), it is *monoclinic*, otherwise *orthorhombic*.

A few examples will serve to illustrate the above rules. The full symmetry group of a cube is m$\bar{3}$m, indicating a threefold rotary-inversion with axis along a main diagonal $\mathbf{a} + \mathbf{b} + \mathbf{c}$, a mirror plane with normal $\mathbf{c}$ and a diagonal mirror plane with normal $\mathbf{a} - \mathbf{b}$; operating with these symmetries on each other produces all the 48 symmetry elements of the cube. The rotational subgroup is 432. The symmetry group of a regular tetrahedron is $\bar{4}$ 3m. The symmetry group of a regular hexagonal prism is 6/mmm; that of a hexagonal antiprism is $\bar{3}$m2.

The H-M nomenclature provides a name for each of the 230 space groups. The first symbol in this H-M name is a capital letter, P, F, I, C or R, referring to the nature of the underlying lattice (i.e. the translational subgroup of the space group). For 71 of the space groups, this is followed simply by the name of a point group. These are the *symmorphic* space groups. The H-M symbol for the remaining space groups follow the same conventions described above concerning the orientations of axes and planes, except that *screw axes* can take the place of rotation axes and *glide planes* can take the place of mirror planes.

A *screw transformation* is a combination of a rotation and a translation along the direction of the axis. Screw axes occurring as elements in space groups are 2_1, 3_1, 3_2, 4_1, 4_2, 4_3, 6_1, 6_2, 6_3, 6_4, 6_5. A screw transformation n_m is the result of the successive application of rotation through $2\pi/n$ and a translation along the axis equal to a fraction m/n of the corresponding primitive lattice translation. A *glide transformation* is the effect of the successive application of reflection in a plane followed by translation parallel to the plane and equal to half of a primitive lattice translation. The letters a, b and c denote glides with translational components $\mathbf{a}/2$, $\mathbf{b}/2$ and $\mathbf{c}/2$, respectively. A glide with a translational component such as $(\mathbf{a} + \mathbf{b})/2$ is denoted by n. The letter d also occasionally appears in an H-M symbol. It refers to a glide whose translational component is half of a centering translation, such as $(\mathbf{a} + \mathbf{b})/4$ in an F or C type group or $(\mathbf{a} + \mathbf{b} + \mathbf{c})/4$ in an I-type group.

All this is less complicated than it seems on a first encounter. An effective way of gaining familiarity with it would be to try to see the various symmetry transformations indicated in the H-M symbol of the structures illustrated in the figures.

3.2 Packing of Regular and Semi-regular Polyhedra

A tiling of 3D space is a packing of polyhedra leaving no gaps. The term 'polyhedron' is very general; a useful beginning is to consider vertex-transitive triply periodic tilings of E_3 by semi-regular polyhedra. Eleven of them are simply constructed from layers of prisms wherein the layers correspond to the patterns of

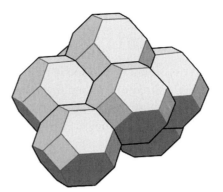

Figure 3.2 Space-filling by truncated octahedra.

Kepler's 2D tilings by regular polygons (Figure 2.1):

Pm $\bar{3}$m:	4^3	P6/m:	3.4^2	P6/mmm:	$4^2.6$
Cmmm:	$4^3 + 3.4^2$	P4/mmm:	$4^3 + 4^2.8$	P4/mbm:	$4^3 + 3.4^2$
P6/m:	$4^2.6 + 3.4^2$	P6/mmm:	$3.4^2 + 4^2.6, 4^3 + 3.4^2 + 4^2.6,$		
			$3.4^2 + 4^2.12, 4^3 + 4^2.6 + 4^2.12$		

The only semi-regular polyhedron that is space-filling is the truncated octahedron, which gives the 3D tiling Im3m: 4.6^2 (Figure 3.2). This configuration is of special interest in the theory of cellular structures (Kelvin & Weaire 1997; Sadoc & Mosseri 1999). Lord Kelvin (Thompson 1887; 1894) conjectured that the cellular structure with the least area per unit volume, subject to the condition that all cells be congruent, is a structure derived from the 4.6^2 packing by minimising the surface energy – the facets of the polyhedral cells are then slightly curved, as are the edges, so that the faces meet at $120°$ and the edges meet at the 'tetrahedral angle' $109.4°$. The truncated octahedron is for this reason sometimes referred to as the 'Kelvin polyhedron'.

Andreini (1907) appears to have been the first to look for further examples of uninodal 3D tilings with regular and semi-regular tiles. He found the following 12, which have been illustrated by various authors (Critchlow 1969; 2000; Pearce 1978; Williams 1979; etc.):

Fd$\bar{3}$m:	$3^3 + 3.6^2$
Fm$\bar{3}$m:	$3.6^2 + 3.8^2 + 4.6.8, 3^3 + 3^4, 3^3 + 4^3 + 3.4^3,$
	$3.6^2 + 3.4.3.4 + 4.6^2$
Pm$\bar{3}$m:	$3^4 + 3.4.3.4, 3^4 + 3.8^2, 4^3 + 3.4^3 + 3.4.3.4,$
	$4^3 + 4^2.8 + 3.8^2 + 3.4^3, 4^3 + 4.6^2 + 4.6.8$
Im$\bar{3}$m:	$4^2.8 + 4.6.8, 4.6^2$
P6$_3$/mmc	$3^3 + 3^4$

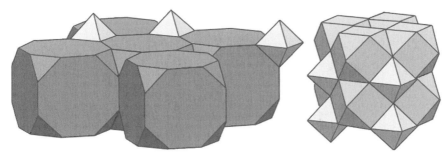

Figure 3.3 The tilings Pm$\overline{3}$m: $3.8^2 + 3^4$ and Pm$\overline{3}$m: $3.4.3.4 + 3^4$.

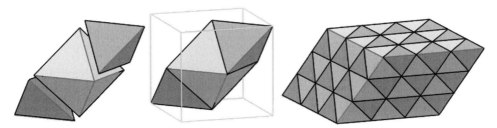

Figure 3.4 The space-filling Fm$\overline{3}$m: $3^3 + 3^4$.

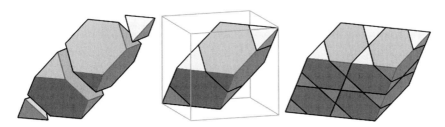

Figure 3.5 Fd$\overline{3}$m: $3^3 + 3.6^2$.

Truncating the cubes of the regular packing of cubes produces voids in the shape of octahedra, so we get a regular space-filling with semi-regular tiles of two kinds, *truncated cubes* and *octahedra*. Further truncation produces a filling by *cubocta-hedra* and *octahedra* (Figure 3.3).

The primitive unit cell of the *face-centred cubic* (fcc) lattice can be dissected into two tetrahedra and an octahedron. When repeated by translations this gives a space-filling of *tetrahedra* and *octahedra* (Figure 3.4).

The primitive unit cell of the fcc lattice can, alternatively, be partitioned into two *tetrahedra* and two *truncated tetrahedra*, and we are led to the polyhedron packing shown in Figure 3.5.

Truncating the octahedra and the tetrahedra of the space-filling Fm$\overline{3}$m: $3^3 + 3^4$ leaves cuboctahedral voids, and we arrive at a space-filling by *truncated tetrahedra*,

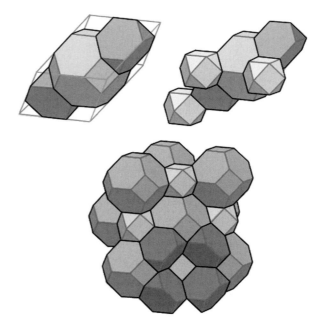

Figure 3.6 Fm$\bar{3}$m: 3.62 + 4.62 + 3.4.3.4.

truncated octahedra and *cuboctahedra* (Figure 3.6). Boron atoms and metal atoms can form a configuration like this, with metal atoms at the centres of the truncated octahedra coordinated to 24 boron atoms at the vertices of the polyhedra.

An array of *cuboctahedra* centred on a primitive cubic lattice can be connected by *cubes* linking their square faces, The voids in this structure can be filled by *rhombicuboctahedra* (3.4^3). Similarly, an array of *truncated octahedra* can be linked by *cubes*, giving a structure with voids that can be filled by *truncated cuboctahedra* (4.6.4.8). We get the two polyhedral packings shown in Figure 3.7. The four connected network of edges of this latter configuration is the zeolite framework LTA.

An array of *truncated cubes* centred on a primitive cubic lattice can be linked by *octagonal prisms*. The remaining voids can then be filled by *rhombicuboctahedra* and *cubes* (Figure 3.8). *Truncated cuboctahedra* in contact on hexagonal faces are centred on a body-centred cubic (bcc) lattice. The voids can be filled by *octagonal prisms* (Figure 3.9).

Figure 3.10 shows an fcc packing of truncated cubes, truncated cuboctahedra and truncated tetrahedra. An fcc array of closely packed *rhombicuboctahedr*a has voids in the shape of *cubes* and *tetrahedra* (Figure 3.11).

The packing of tetrahedra and octahedra, whose vertices constitute an fcc lattice, can be built up in layers. Changing the stacking sequence of the layers changes the

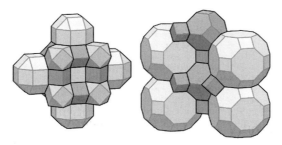

Figure 3.7 The space-fillings Pm$\overline{3}$m: $4^3 + 3.4^3 + 3.4.3.4$ and Pm$\overline{3}$m: $4^3 + 4.6^2 + 4.6.8$.

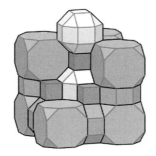

Figure 3.8 Fm$\overline{3}$m: $4^3 + 4^2.8 + 3.8^2 + 3.4^3$.

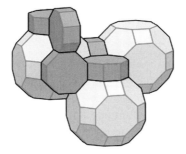

Figure 3.9 Fm$\overline{3}$m: $4^2.8 + 4.6.8$.

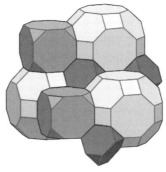

Figure 3.10 Fm$\overline{3}$m: $3.6^2 + 3.8^2 + 4.6.8$.

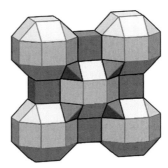

Figure 3.11 Fm$\bar{3}$m: $3^3 + 4^3 + 3.4^3$.

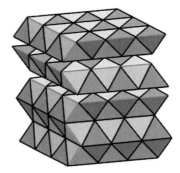

Figure 3.12 6_3/mmc: $3^3 + 3^4$.

symmetry of the arrangement to a hexagonal symmetry. We get P6$_3$/mmc: $3^3 + 3^4$, Figure 3.12.

This completes Andreini's list of uninodal tilings of 3D space by regular and semi-regular polyhedra. There are, in fact, four other possibilities (Grünbaum 1994), which we list below and illustrate in Figure 3.13:

P6$_3$/mmc:	$3^3 + 3^4 + 3.4^2$	R$\bar{3}$m:	$3^3 + 3^4 + 3.4^2$
I4$_1$/amd:	$3.4^2, 3.4^2 + 4^3$	I4$_1$/amd:	$4^3 + 3.4^2$

Layers of triangular prisms can alternate with the layers of octahedra and tetra-hedra, without disturbing the *uninodal* property of the structure. We get (3.13a) P6$_3$/mmc: $3^3 + 3^4 + 3.4^2$ and (3.13b) R$\bar{3}$m: $3^3 + 3^4 + 3.4^2$. The packing of triangular prisms P6/m: 3.4^2 can be changed by rotating alternate layers through 90° to give (3.13c) I4$_1$/amd: 3.4^2. Finally, layers of *cubes* can alternate with the layers of triangular prisms, giving (3.13d) I4$_1$/amd: $4^3 + 3.4^2$.

Many of these polyhedral packings can be regarded as *modules* from which descriptions of crystal structures can be built – for example by placing atoms at the

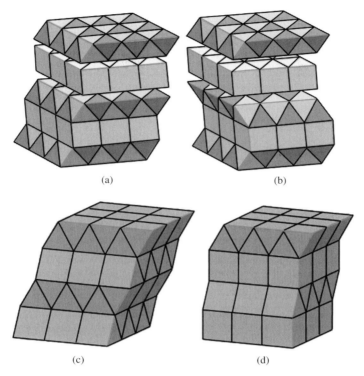

Figure 3.13 (a) P6$_3$/mmc: $3^3 + 3^4 + 3.4^2$; (b) R$\bar{3}$m: $3^3 + 3^4 + 3.4^2$; (c) I4$_1$/amd: 3.4^2; (d) I4$_1$/amd: $4^3 + 3.4^2$.

centre of a polyhedron and atoms of the same or a different species at the vertices (the polyhedron is a *coordination polyhedron* for its central atom).

Here are a few examples.

In the mineral *perovskite* CaTiO$_3$ and minerals of the perovskite type ABO$_3$, oxygen atoms occupy the vertices of the polyhedra in the packing of cuboctahedra and octahedra (Figure 3.3b), and atoms of two kinds, A and B, occupy the centres of the cuboctahedra and octahedra, respectively.

The structure of the silicate *β-crystobalite* can be derived from Figure 3.5 (Fd$\bar{3}$m: $3^3 + 3.6^2$) by interpreting the tetrahedra as SiO$_4$ tetrahedra – silicon atoms are located at the centres of the tetrahedra and all the vertices of the polyhedra are sites for the oxygen atoms. The *Friauf–Laves phases* of intermetallics can be described in terms of the same polyhedron packing, the vertices of the polyhedra are sites for one type of atom, and larger atoms are located at the centres of the truncated tetrahedra. The truncated tetrahedron is also sometimes referred to as the *Friauf polyhedron*.

The packing shown in Figure 3.6 corresponds to a complex arrangement involving boron and metallic atoms. The vertices are sites of boron atoms, the truncated octahedra are coordination shells surrounding metal atoms, each coordinated to 24 boron atoms. Note how the metallic atoms are linked to each other through cuboctahedral boron clusters. Other quite different structures can arise depending on the proportion of boron to metal; they are also readily describable in terms of other polyhedral space-fillings (Sullinger & Kennard 1966).

A classification of types of intermetallic structure in terms of coordination polyhedra and their methods of combination has been presented by Krypyakevich (1963).

3.3 Voronoi Regions

The *Voronoi region* of a point p of a set of points is the polyhedron containing all locations that are closer to p than to any other point of the set (Voronoi 1908); if the point set is the set of positions of atoms in a crystalline structure the vertices of the Voronoi polyhedra are voids in the structure that can be occupied by atoms of a different type. The corresponding 2D concept is a *Dirichlet region* (Dirichlet 1850).

Since the Andreini type polyhedral packings are vertex-transitive, their *dual* configurations (whose vertices are at the centres of the polyhedral tiles) are *tile*-transitive. That is, they are tilings of E_3 by *congruent* polyhedra. The dual of Fm$\overline{3}$m: $3^3 + 3^4$, for example, is the space-filling by rhombic dodecahedra. The polyhedral tiles of the dual configurations are the Voronoi regions of the vertices.

Plate III shows the 3D tiling by the Voronoi regions of the atoms in the complex intermetallic Mg_{32} (Al, Zn)$_{49}$, which will be described in Chapter 6.

Cornelli *et al.* (1984) and Loeb (1962; 1963; 1970; 1974; 1990) have explored this idea and described efficient algorithms for storing and generating the structures of minerals and alloys. Blatov and Shevchenko (2003) have developed computer

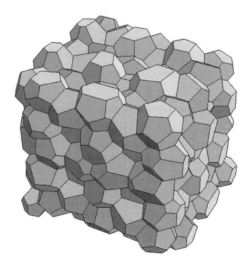

Plate III The Voronoi regions around the 152 atoms in a unit cell of $Mg_{32}(Al, Zn)_{49}$.

algorithms for estimating void sizes and visualising the system of interconnecting channels between voids. The method proceeds by first finding the Voronoi dissection associated with the positions of the atoms, and then a further Voronoi dissection is determined from the point set given by the atomic positions *and* the vertices of the original Voronoi dissection. We then obtain two networks of polyhedra, one composed of the polyhedra centred around the atoms and one composed of the polyhedra surrounding the vertices. This second polyhedral network represents the void and channel structure. An interesting feature is the way in which sets of close vertices of the original dissection conglomerate to form a single void.

3.4 Fedorov's Parallelohedra

Fedorov (1891) showed that there are just five topological types of convex polyhedra that can tile E_3, if we require that all tiles be identical and all in the same orientation. The faces of such a polyhedral tile must occur in pairs of congruent, parallel faces – Fedorov's tiles are *parallelohedra*, which are special kinds of *zonohedra* (Coxeter 1963). The Voronoi regions of lattices are Fedorov polyhedra, as are variants obtained from them by affine deformations (uniform contraction or expansion, and shear), or by lengthening or shortening of a set of parallel edges. In their highest symmetry versions, the five types are: cube, hexagonal prism, truncated octahedron, rhombic dodecahedron and a polyhedron (Figure 3.14) with four hexagonal and four rhombic faces.

Delone has classified 3D lattices in terms of the type of Fedorov zonohedron that occurs as the Voronoi region of a lattice point. Two lattices of the same Bravais type may have different kinds of Voronoi region, depending on the c/a ratio. There are 14 3D lattices according to the Bravais classification and 24 according to Delone's scheme (Delone *et al.* 1934).

3.5 Lattice Complexes

A *lattice complex* is a set of special positions on which a space group acts transitively (i.e. any point of the set can be mapped to any other by the action of the group); those

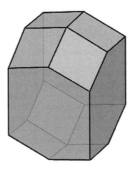

Figure 3.14 One of Fedorov's parallelohedra.

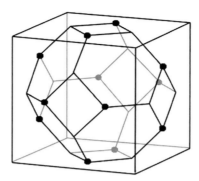

Figure 3.15 The vertices of the truncated octahedron that define the W complex. The bounding cube is a unit cell of Pm$\overline{3}$.

that are observed to occur most frequently as atomic positions in crystalline structures have been named (Hellner 1965; Fischer *et al.* 1973; Fischer 1991; 1993; Fischer & Koch 1995). About 2000 crystal structures can be generated from only 10 lattice complexes (Loeb 1970). For example, the vertices of the space-filling Im$\overline{3}$m: 4^3 of cubes constitute the primitive cubic lattice, which is the lattice complex P in its highest symmetry form. The complex I is P + P′ where P′ denotes P shifted by (½, ½, ½), F is the vertex set of Fm$\overline{3}$m: $3^3 + 3^4$, and D is F + F′ where F′ is F translated by (¼, ¼, ¼). The complex J is the vertex set of Pm$\overline{3}$m: $3^4 + 3.4.3.4$ (octahedra and cuboctahedra) and T is the vertex set of Fd$\overline{3}$m: $3^3 + 3.6^2$ (tetrahedra and truncated tetrahedra). The complex W consists of 50% of the vertices of the space-filling by truncated octahedra, Im$\overline{3}$m: 4.6^2, arranged as shown in Figure 3.15. The symmetry is reduced to Pm $\overline{3}$. W + I specifies the atomic positions in β-tungsten (β-W).

By allowing polyhedra with slight metrical deviations from regularity, further vertex-transitive polyhedron packings can be produced. (A surprising variety of monohedral tilings of E$_3$ exists, in which the polyhedra are *topologically* equivalent to the Platonic solids (Delgado-Friedrichs & Huson 1999); if they are required to be *metrically* the Platonic solids, then, of course, only the cube tiles E$_3$.) This is an important generalisation because atoms are not hard spheres, and bond length and bond angles are, within limits, flexible, so that nature is able to produce some surprising structures. Observe, for example, that the positions indicated in Figure 3.15 lie at the vertices of a distorted *icosahedron* (Figure 3.16), and we get a vertex-transitive packing of icosahedra and tetrahedra ($3^5 + 3^3$). In the β-W structure, atoms are also located at the bcc positions, that is at the centres of the icosahedra. This was earlier called 'beta-tungsten' but was actually W$_3$O. Other examples of the occurrence of the β-W configuration are Cr$_3$Si, Nb$_3$Sn, V$_3$Ga and V$_3$Ge. Partitioning each icosahedron into 12 tetrahedra, we obtain a space-filling by tetrahedra. The structure is binodal, with 12-coordinated and 14-coordinated atoms. β-W is one of the

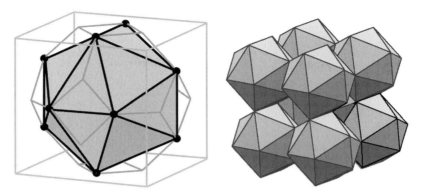

Figure 3.16 The β-W structure: (a) 12 of the 24 vertices of a truncated octahedron define a slightly irregular icosahedron and (b) the voids in the array of icosahedra are irregular tetrahedra. Note that the icosahedron in the centre of this figure is oriented differently from the surrounding eight, so the structure is not actually bcc.

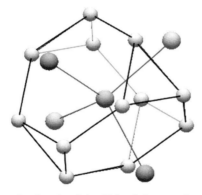

Figure 3.17 A structural subunit of the Friauf–Laves phase. The central atom is coordinated to 12 atoms at the vertices of a truncated tetrahedron and to four at the centres of neighbouring truncated tetrahedra.

simplest of the Frank–Kasper phases (Frank & Kasper 1958). Another simple (and important) example is the Laves phase, in which atoms occur at the vertices and centres of the truncated tetrahedra, in the space-filling Fd$\bar{3}$m: $3^3 + 3.6^2$. The atoms at the centres of truncated tetrahedra are 16-coordinated (Figure 3.17), those at the vertices are 12-coordinated.

3.6 Polytetrahedral Structures

Regular tetrahedra, of course, will not tile E$_3$. The dihedral angle of a regular tetrahedron is 70.53°, so that five tetrahedra around an edge will leave five angular gaps

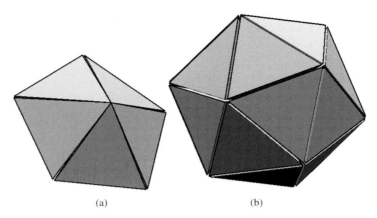

(a)   (b)

Figure 3.18 (a) Five regular tetrahedra sharing a common edge and (b) 20 regular tetrahedra sharing a common vertex.

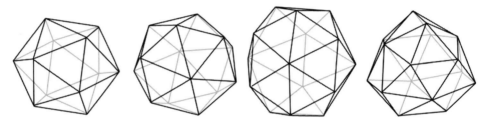

Figure 3.19 Coordination shells corresponding to coordination numbers 12, 14, 15 and 16, in Frank–Kasper phases.

of only 1.47° between faces, as in Figure 3.18a; these gaps can be closed by very small deviations of the tetrahedra from perfect regularity. Similarly, there are small gaps (about 2.9° between the faces that are nearly in contact) when 20 tetrahedra share a common vertex, as in Figure 3.18b; a slight deformation will produce a regular icosahedron. Somewhat greater geometric frustration is involved when six tetrahedra share a common edge. Frank and Kasper showed how the structures of many quite complex alloys, with every atom coordinated to 12, 14, 15 or 16 atoms, can be understood in these terms (see Figure 3.19). The Frank–Kasper phases are *polytetrahedral structures*. The atoms lie at the vertices of a space-filling by 'almost regular' tetrahedra. An edge shared by six tetrahedra rather than five is a *disclination line*. These disclination lines form a network that characterises the particular structure. Atoms at the centres of the 14-, 15- or 16-coordination shells are nodes of the disclination network, with connectivities 2, 3 or 4, respectively. The Voronoi regions surrounding the 12-, 14-, 15- and 16-coordinated atoms in a Frank–Kasper phase are 12-, 14-, 15- and 16-hedra, with 12 pentagonal faces and 0, 2, 3 or 4 hexagonal faces, respectively (Figure 3.20). The edges of the disclination net cut through the hexagons.

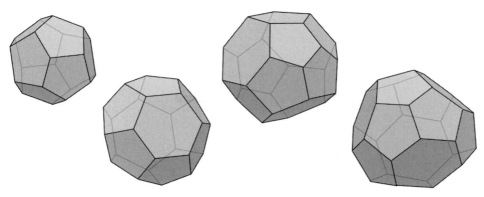

Figure 3.20 Voronoi regions for Frank–Kasper phases.

Polytetrahedral order is of considerable importance, not only for the crystalline Frank–Kasper phases. Polytetrahedral packings play a role in the theories of liquids and glasses – see the extensive review by Nelson & Spaepen (1989) – and in the structure of small clusters of atoms and molecules.

3.7 The Polytope {3, 3, 5}

Perfectly regular tetrahedra can be packed together in a spherical space S_3. On a hypersphere embedded in Euclidean space E_4 the vertices are those of the regular polytope $\{3, 3, 5\}$. This polytope has 120 vertices, 720 edges, 1200 equilateral triangular faces and 600 regular tetrahedral cells (Coxeter 1963). There are five tetrahedra around every edge and 12 around every vertex (forming a regular icosahedron). Sadoc and Mosseri (1999) have investigated the disclination nets of Frank-Kasper structures quite extensively. They describe the structures in E_3 in terms of unfolding of the E_4 polytope $\{3, 3, 5\}$.

3.8 Packings Involving Pentagonal Dodecahedra

The Voronoi regions for Frank–Kasper phases are polyhedra with 12, 14, 15 or 16 faces (12 pentagons and 0, 2, 3 or 4 hexagons, respectively). The arrangement of atoms in these phases can be visualised in terms of space-fillings of these polyhedra. The dihedral angle of a regular pentagonal dodecahedron is 116.6°, so that three regular dodecahedra around a common edge leave a defect angle of only 3.4° between faces. Hence, three very slightly irregular pentagonal dodecahedra can share an edge, while each pair has a common face. Indeed, a cluster of four face-sharing pentagonal dodecahedra, in a tetrahedral configuration, is possible (Figure 3.21). These clusters can be packed to produce a diamond-type network; the voids constitute another

Figure 3.21 A cluster of four face-sharing almost regular dodecahedra.

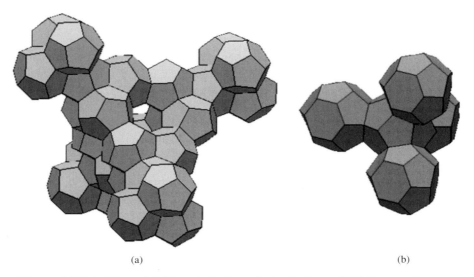

(a)   (b)

Figure 3.22 (a) Extended D-net built from dodecahedra and (b) the complementary D-net of 16-hedra.

diamond-type network consisting of the 16-hedra (Figure 3.22). This packing of dodecahedra and 16-hedra corresponds to the partitioning of space by the Voronoi regions of the Laves phase. Pearce (1978) and Williams (1979) have illustrated and described in detail other ways in which the 12, 14, 15 and 16 faced polyhedra can be packed in space-filling arrangements. The configuration shown in Figure 3.23, built from dodecahedra and 14-hedra, is of special interest. It can be visualised as a packing of rods formed as stacks of 14-hedra, the rods oriented in three mutually perpendicular directions. The voids in this structure are pentagonal dodecahedra, centred on a bcc lattice. The edges of this polyhedral packing describe the network in chlorine hydrate – whose structure was solved by Pauling & Marsh (1952) – an example of a

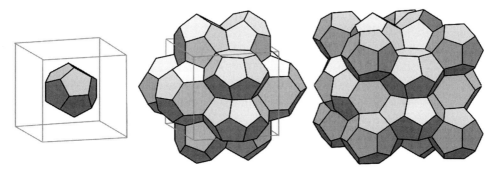

Figure 3.23 Twelve 14-hedra packed around a pentagonal dodecahedron. The structure can be extended indefinitely, with dodecahedron centres at bcc positions. The symmetry group is Pm$\bar{3}$m (not Im$\bar{3}$ because the central dodecahedron and those at the corners of the unit cube are differently oriented. This cellular structure is the Voronoi region structure of β-W.

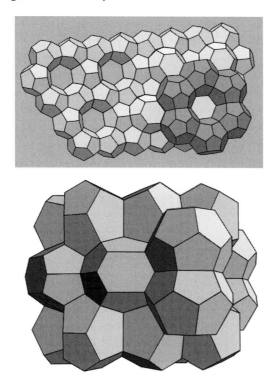

Plate IV Two views of the dissection of space into Voronoi regions of Zr_4Al_3.

clathrate structure (see Chapter 8). This packing of dodecahedra and 14-hedra also corresponds to the Voronoi regions for β-W. Weaire and Phelan discovered that when the surface areas of this cellular structure are minimised subject to the condition that all the cells shall have equal volume, an idealised foam is obtained with an area per unit volume some 3% less that that of Kelvin's structure (Weaire & Phelan 1996; Rivier 1996). (This has been frequently described as a 'disproof of Kelvin's conjecture', which it is not. Kelvin's conjecture was about *congruent* cells.)

Plate IV shows the Voronoi regions for the hexagonal Frank–Kasper phase Zr_4Al_3 (Wilson *et al.* 1960) – a tiling of space by the 12-, 14- and 15-hedra.

3.9 Asymmetric Units

The *International Tables for Crystallography* (Hahn 1995) lists, for each of the 230 space groups, the vertices of an 'asymmetric unit' or 'fundamental region' associated with the group. It is a polyhedron that produces a tiling of E_3 when the group operations are applied, and that does not itself have any symmetry belonging to the group. All the possible shapes for the units associated with the cubic space groups were derived by Koch and Fischer. Their method was based on the partitioning of the Voronoi region of a point set generated by the group (Koch and Fischer 1974). Any triply periodic structure in E_3 is completely specified by a description of the portion of it contained in an asymmetric unit – the units are thus modules for the building of triply periodic structures. Two contiguous units in the tiling are related to each other by a group operation. This set of transformations associated with a unit will generate the whole space group (Lord 1997); repeated application of them will generate any triply periodic structure from the portion of it that lies within the unit. The simple example given in Figure 3.24 should suffice to indicate the general idea: an asymmetric unit of the space group $Im\overline{3}m$ is a tetrahedral tile, related to neighbouring units by a diad rotation or by reflection. There are 96 such units per unit cell of $Im\overline{3}m$. The tetrahedron formed from two of these units related by diad

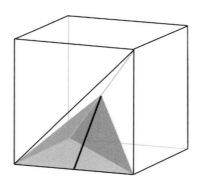

Figure 3.24 Example of an 'asymmetric unit'. The shaded tetrahedron is an asymmetric unit for the space group $Im\overline{3}m$. The heavy line on the front face is a twofold rotation axis and the other three faces are reflection planes. The cube is an eighth of a unit cell.

rotation has all faces mirror planes. This is the asymmetric unit of Pm$\overline{3}$m – a 'group generated by reflections' or *Coxeter group*.

An interesting recent development has been the study of the asymmetric units considered as topological spaces. Essentially, the idea is to identify pairs of facets of the unit that are mapped to each other by a transformation belonging to the space group. The resulting topological object is an *orbifold*. Each space group is associated with a *unique* orbifold (whereas for most of the space groups there is arbitrariness in the choice of asymmetric unit) (Delgado-Friedrichs & Huson 1997; Johnson *et al.* 2002).

3.10 Modular Structure

In Section 3.2, we drew attention to the possibility of a modular description of crystalline materials in which the atoms are located at vertices and centres of the polyhedral tiles of a tiling of E$_3$. The polyhedral tiles are *modules* from which a visualisation of the structure can be assembled. The tiling of space by the asymmetric units of a space group provide another example of the modular approach to structure description. All the tiles are identical and contain identical portions of any triply periodic structure with that space group symmetry.

The description of triply periodic structures in terms of 3D tilings is rich in possibilities. We illustrate the method through the introduction of two particularly interesting examples: the description of triply periodic surfaces, and the description of 3D weaving patterns.

Consider the surface patch inside a cubic module, shown in Figure 3.25. A *continuous* surface can be produced by assembling these cubic modules to form the tiling Im$\overline{3}$m: 4^3, provided we ensure that the surface matches across the boundary between two cubes. Since the surface patch within the module can have four distinct orientations, an infinite number of topologically distinct surfaces can be generated. One such possibility is shown in Figure 3.26. Observe that each face of the cube cuts the surface in a pair of arcs like those decorating Cyril Smith's version of the Truchet tilings (Figure 2.8c). We have, therefore, a 3D analogue of these Truchet tilings. Note that the relative orientation of neighbouring tiles is

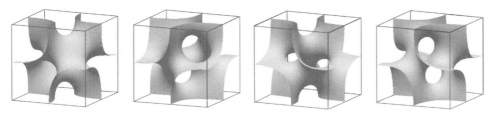

Figure 3.25 Four orientations for a surface patch in a cubic module.

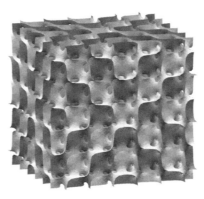

Figure 3.26 A unit cell of a triply periodic surface containing sixty-four cubic modules.

restricted by the requirement that the edges of the surface patches have to match across the tile boundaries, whereas for the 2D tilings there was no such restriction.

In Chapter 9, we shall discuss various kinds of triply-periodic *minimal* surfaces. The unit shown in Figure 3.25 has the form of one-eighth of a unit cell of the minimal surface known as 'Schoen's batwing' (see Ken Brakke's website, and Lord & Mackay 2003). Neighbouring units are related everywhere by reflection in their common face. For the minimal surface named by Brakke the 'pseudobatwing', a similar module is the unit cell – all the modules have the same orientation. The batwing has symmetry Im$\bar{3}$m and the pseudobatwing has rhombohedral symmetry R$\bar{3}$m. It is quite remarkable that the difference in shape of the modular surface patches for these two distinct minimal surfaces is imperceptible. This suggests the possibility of an infinite number of possible minimal surfaces, based on the 3D Truchet tiling principle.

As a further example of the application of modularity to the description of 3D structures, we shall briefly examine some possibilities for 3D weaving patterns. The simplest 2D weave has two orthogonal thread directions (warp and weft) and a simple under-over-under-over- … structure. 3D generalisations have recently become of increasing importance in the manufacture of composite materials, in which fibres embedded in a matrix constitute a reinforcement of the material. Derck van Schuylenburch has investigated the possibilities for 3D weaving patterns, with three orthogonal thread directions, subject to the condition that every 2D layer is a simple 2D under-over-under-over- weave (a 'mat'). He holds patent rights to these 3D weaving methods. We introduce here a modular approach to describing the possibilities. A cubic module containing portions of three fibres, as in Figure 3.27, can be thought of as a tile from which various 3D weaves can be generated. Contiguous tiles can be related by rotation about an axis in a cube face, by screw transformations, by glides, reflections, etc. However, the requirement that the resulting pattern shall consist of

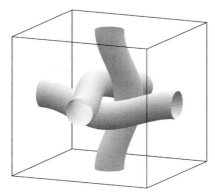

Figure 3.27 A module for generating 3D weaving patterns.

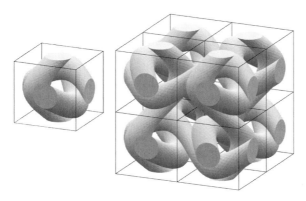

Figure 3.28 A module (one-eighth of a unit cell) and a unit cell of OOO, with sinusoidal fibres.

stacks of simple 2D mats, in all three directions, is severely restrictive. Van Schuylenburch discovered that there are then just five possibilities. Indeed, neighbouring cubes can be related only by inversion in their common face or by twofold screw transformations (respectively O and I in van Schuylenburch's notation). In terms of a three-letter symbol listing the kind of transformation to be applied along the three orthogonal directions, we have get four weaving patterns

OOO, OOI, OII, III

The symmetries of these patterns are, respectively:

R$\bar{3}$c, Pnc2, Iba2, I23

(III exists in a right-handed form IIIR or a left-handed form IIIL, according to whether the basic module is the one shown in Figure 3.27 or its mirror image.) We illustrate the two varieties with cubic symmetries, in Figures 3.28 and 3.29.

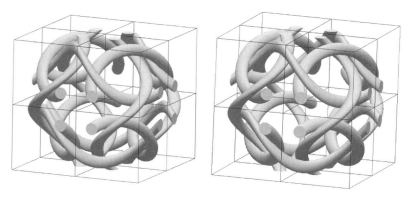

Figure 3.29 A pair of stereo images of a unit cell of the weave IIIR, constructed from helical fibres.

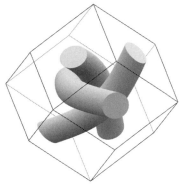

Figure 3.30 A module for the generation of the 3D kagome weave.

The rhombic dodecahedron module shown in Figure 3.30 produces a 3D weaving pattern with symmetry $F4_132$ that has *six* thread directions corresponding to the edges of a regular tetrahedron. Only half of the rhombic dodecahedra in the polyhedral packing contain portions of the pattern, the other dodecahedra in the space-filling are empty. The resulting pattern (Figure 3.31) consists of *four* stacks of 2D mats parallel to the four tetrahedron faces; each mat is a simple kagome type basket weave.

We shall encounter the modular principle again in subsequent chapters. In particular cases the objects within a module may be sets of points (0D) as in the specification of atomic positions, they may be 1D (fibres) as in the 3D generalisation of weaving, braiding, knitting (and rod-packings as a very special case), 2D (surfaces), or 3D, as for example an electron density distribution. The 1D case has been least explored. In the examples we have shown the fibres are infinite in length. It is also possible for fibres in a pattern to take the form of closed loops intertwined in

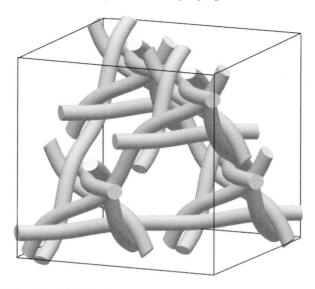

Figure 3.31 A unit cell of the 3D kagome weave.

various ways. Clearly, the modular approach to structure opens up an extensive, largely unexplored territory of geometrical concepts.

3.11 Aperiodic Tilings of E_3

An icosahedron has 6 fivefold axes. A hexahedron with rhombic faces is obtained by choosing its edges along the fivefold directions. The face diagonals are then in the ratio $1 : \tau$, where τ is the 'golden number' $(1 + \sqrt 5)/2$. There are two varieties: a prolate hexahedron with three acute face angles around a vertex, and an oblate hexahedron with three obtuse face angles around a vertex. That these two units are 3D analogues of the Penrose rhombs was first recognised by Robert Ammann in 1976 (Mackay 1981; Gardner 1995). In every tiling by these units, the edges of the units are oriented along the fivefold symmetry axes of an icosahedron; their threefold axes lie along the icosahedral threefold axes, and their face diagonals lie along icosahedral twofold axes. The resulting tilings of E_3 thus have long range orientational order with icosahedral symmetry. Ammann showed that aperiodicity can be imposed by matching rules (see for example Gardner (1995) or the markings devised by D. Levine and reproduced in Steinhardt & Ostlund (1987) and Katz (1986)). Aperiodic tilings with long-range orientational order, built from these units, can be generated by projection from a 6D hypercubic lattice (Kramer & Neri 1984); in this approach, the six fivefold symmetry axes of an icosahedron, corresponding to the six directions of the edges of the 3D pattern, are projections on to E_3 of the edges of a 6D hypercube.

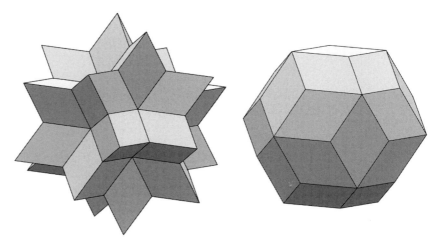

Figure 3.32 The star polyhedron and the rhombic triacontahedron.

Aperiodic patterns in 3D can be obtained from decorations of the two units –
similarly to the way triply periodic patterns can be generated by a decoration of the
asymmetric unit of a space group. An interesting proposal is the 3D generalisation
of the weaving pattern shown in Figure 2.10. The fibres connect the three pairs of
parallel opposite faces of every unit. Imposing a rule for the way the fibres have to
cross in the middle without intersecting, one obtains a 3D weave pattern with 15
sets of fibres running parallel to the 15 twofold axes of an icosahedron. Such a
weave, if it could be implemented in fibreglass or carbon-fibre reinforced plastic,
would be structurally isotropic (Mackay 1988).

A characteristic of the 3D Penrose tilings is the ubiquitous occurrence of *star
polyhedra* consisting of 20 prolate units, and *rhombic triacontahedra* (Figure 3.32).
Kowalewski (1938) had shown that the rhombic triacontahedron can be decomposed
into 10 prolate and 10 oblate units. A demonstration of this is indicated in Figure
3.33. (There are in fact two different ways of constructing a rhombic triacontahedron
from 10 oblate and 10 prolate units. The arrangement shown in the figure is the one
that occurs in the standard 3D Penrose tiling. The other has a threefold symmetry
axis (Longuet-Higgins 1992; 2003).) The two rhombic hexahedra may be conveni-
ently named '*Kowalewski units*'. Longuet-Higgins (2003) has demonstrated the
remarkable fact that a sequence of rhombic triacontahedra, with edge lengths 1, 2,
3,... can be built up in layers from the two units.

Much of the early theoretical work on icosahedral quasicrystals was formulated
in terms of these patterns (see the collection of reprints in Steinhardt & Ostlund 1987;
Mikalkovic *et al.* 1996). Quasicrystals occur in conjunction with closely related
crystalline (i.e. periodic) structures known as *approximants*. Knowledge of the
atomic arrangements in approximants is a valuable guide in attempts to deduce

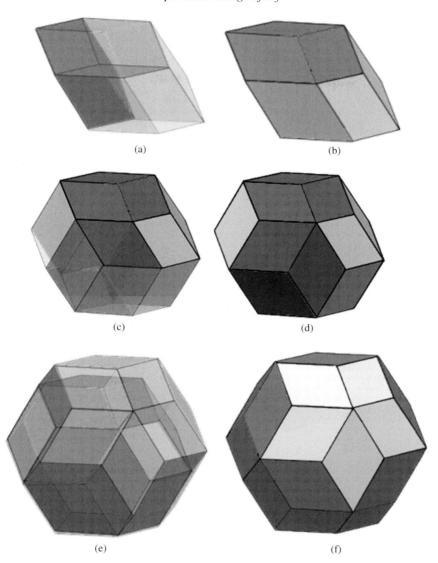

Figure 3.33 The building of a rhombic triacontahedron from 10 prolate and 10 oblate rhombohedra. (a,b) A *rhombic dodecahedron* built from two prolate units and two oblate units. This is the 'rhombic dodecahedron of the second kind' discovered by Bilinski (1960). (c,d) Three prolate and three oblate units added to this gives (d) a *rhombic icosahedron*. (e,f) Finally the rhombic icosahedron is capped by five prolate and five oblate units to give the *rhombic triacontahedron*.

the structure of the related quasicrystal. An interesting and classic example is the model of the crystalline phase α-Al–Mn–Si described by Elser & Henley (1985), which led to insight into the structure of the related icosahedral quasicrystalline phase. In this work, α-Al–Mn–Si is described in terms of a regular packing of two

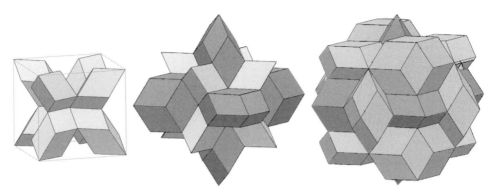

Plate V A periodic packing of prolate Kowalewski units and rhombic dodecahedra; (a) the cubic unit cell contains eight prolate units; (b) the rhombic dodecahedra are centered at mid-ponts of faces of the unit cell and (c) at mid points of the
edges of the unit cell.

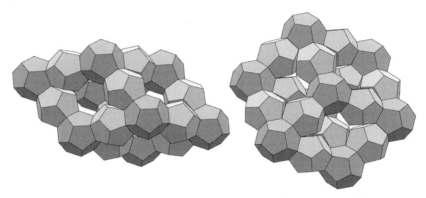

Figure 3.34 The prolate and oblate rhombohedral units constructed from *regular*
dodecahedra (Miyazaki & Yamagiwa 1977). Thus a quasiperiodic network of
face-sharing dodecahedra can be constructed, the centres of the dodecahedra lying
at vertices and mid points of edges of the Kowalewski units. Though it was not
pointed out by these authors, it can be seen that the small angular gaps can be
closed by small deformations.

polyhedral subunits: the prolate Kowalewski units and the rhombic dodecahedron
(Plate V).

Very much before the renewed active interest in the Kowalewski units, that the dis-
covery of quasicrystals aroused, Miyazaki had explored some very remarkable struc-
tures based on the two rhombohedral units (which he named 'golden isozonohedra')
(Miyazaki 1977a, b; Miyazaki & Yamagiwa 1977; Miyazaki & Takada 1980;
Miyazaki 1983; 1986). A striking example of Miyazaki's inventiveness is shown in
Figure 3.34.

4

Circle and Sphere Packing

4.1 Circle Packing

In the Euclidean plane six equal circles can be placed in contact with a circle of the same radius, each touching two of the others. It follows from this that the densest packing of circles in a plane is an array of circles centred at the points of a hexagonal lattice. Kepler's suggestion (Kepler 1611) that the hexagonal forms of snowflakes could be due to the packing of subunits appears to be the earliest insight into the geometrical principles underlying the shapes of crystals.

A *packing* of equal circles may be defined as a configuration of non-intersecting circles in which each touches at least three others. Every such packing can be associated with a *tiling* whose vertices are the circle centres and whose edges join the centres of all pairs of touching circles. Conversely, every tiling pattern with the property that all edges are straight and equal and all angles at tile vertices are not less than 60°, specifies a circle packing. For example, to each of Kepler's 11 uninodal tilings (Figure 2.1) there corresponds a circle packing.

If the centres of all the circles touching a given circle do not all lie on the same side of a diameter of the given circle, the given circle cannot be translated from its position without disturbing other circles – it is *jammed* (Torquato & Stillinger 2001). If every circle in a packing is jammed, the pattern as a whole is *locally jammed*. Interesting cases arise in which a packing is locally jammed, but for which unjamming can take place by the cooperative motion of a group of circles or of the whole configuration. Torquato and Stillinger have provided some interesting insights into this question. They consider, in particular, portions of Kepler-type packings enclosed in a rigid bounding box (the proviso that each circle must touch at least three others is replaced by the requirement that each has at least three contacts – with other circles or with the boundary). Two illustrations give a general idea of the concept. Observe that the packing (Figure 4.1a) can be unjammed by rotating the central group of six circles and that, similarly, the packing (Figure 4.1b) can be unjammed by rotating the central group of four.

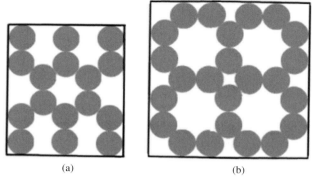

Figure 4.1 Two simple examples of packings that are locally jammed but not collectively jammed.

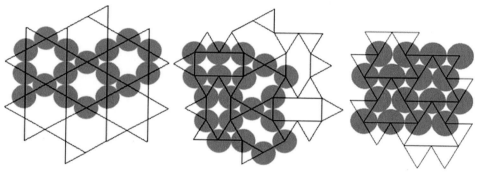

Figure 4.2 A packing of circles in the plane based on the 'Kagome' net, and two distorted versions that it can collapse to by a continuous transformation.

Examples of the collapse of *unbounded* locally jammed packings to packings of higher density by a collective motion are indicated in Figures 4.2 and 4.3. The detailed survey of two-dimensional (2D) nets in crystal chemistry given by O'Keeffe & Hyde (1980) contains various examples of similar transitions relating otherwise apparently unrelated nets. Clearly, similar considerations can be applied to the study of the stability of sphere packings in three or more dimensions.

A curious mathematical problem arises from the question of packing equal circles inside a circular boundary: how must N non-overlapping equal circles be packed inside a unit circle so that their diameter is a large as possible? This question has some practical application in the design of ropes and cables. Tarnai (1998; 2000) has provided an interesting review of the literature on this problem. Solutions for small N are to be seen in nineteenth century Japanese geometry books – the geometrical problem was presumably inspired by the packings of cylindrical bundles of bamboo sticks. The question is equivalent to the question of how N points should be distributed inside a

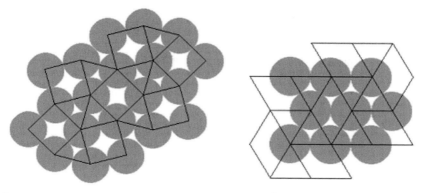

Figure 4.3 The circle packing base on Kepler's $3^2.4.3.4$ net can collapse by a continuous transformation to the densest packing 3^6.

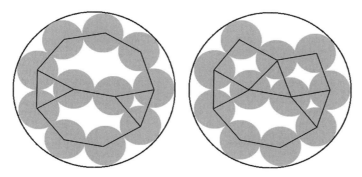

Figure 4.4 The two solutions for the densest packing of 11 circles in a circle.

circle so as to maximise the minimum distance between any two of them. Tarnai has noticed that the packing of lotus seeds in their capsules follow these dense-packed arrangements. Figure 4.4 shows the two optimal solutions that exist for $N = 11$. For $N = 18$ there are 10 configurations.

4.2 Cubic Close Packing

In the cubic close packing (ccp) of equal spheres the sphere centres lie at the vertices of a face-centred cubic (fcc) lattice. Every sphere is in contact with 12 others. The *density*, or *packing fraction* (fraction of space occupied by the spheres) is $\pi/\sqrt{18} = 0.7405\ldots$ Kepler (1619) conjectured that ccp has the highest possible density for a sphere packing in E_3. A proof was obtained only recently (Hales 1997; Sloane 1998; Aste & Weaire 2000; Hsiang & Hsiang 2002). The marked contrast between the difficulty of obtaining a formal proof and the 'self-evident' nature of the analogous 2D problem reflects the fact that, whereas equilateral triangles can tile the Euclidean plane, regular tetrahedra do not tile 3D-Euclidean space.

4.3 Sphere Packings and Nets

Sphere packing problems can, in an obvious way, be treated as problems about 3D nets – the centres of the spheres are the vertices and lines joining the centres of spheres in contact are the edges of a network. Packings of equal spheres are thus equivalent to nets with the special properties that all edges are of equal length and that the angle between two edges sharing a vertex is not less than 60°. Particular cases of nets with these properties are given by the sets of vertices and edges of the space fillings by semi-regular polyhedra (e.g. ccp corresponds to the packing $Fm\bar{3}m$: $3^3 + 3^4$ of tetrahedra and octahedra).

4.4 Low Density Sphere Packings

Heesch & Laves (1933) appear to have been the first investigators to consider sphere packings of exceptionally *low* density. They considered the possibilities for uni-nodal 3- and 4-connected nets, and imposed an additional 'rigidity' requirement that the three, or four, spheres in contact with a generic sphere should not have their centres all on one side of a diametral plane. They described two 3-connected configurations and four 4-connected configurations.

The net for Heesch and Laves' sphere-packing 3_1 (Figure 4.5) is remarkable for several reasons. It is completely regular in the sense that its symmetry group ($I4_132$) acts transitively on the vertices, on the edges and on the rings (shortest circuits through a pair of edges with a common vertex). The rings are decagons (Figure 4.5).

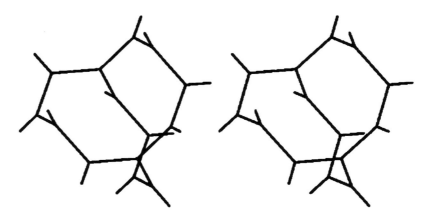

Figure 4.5 Stereo pair of images of a portion of the net for the Heesch-Laves sphere packing 3_1. Note the three decagonal rings.

The net is Wells' (10, 3)-a (Wells 1977). It is *chiral* – existing in two enantiomorphic versions, which in fact are labyrinth graphs for the gyroid surface (see Chapter 9). It is sometimes named the $SrSi_2$ net because the silicon atoms in $SrSi_2$ are at the vertices of this net.

The sphere packing with the least density is Heesch and Laves' 3_2 (Figure 4.6), with a density 0.0555. It can be obtained from 3_1 by replacing every sphere by a group of three spheres. Equivalently, its net is a 'truncated' 3_1 net; it consists of 3-rings and 20-rings.

4_1 is simply the well-known arrangement of carbon atoms in diamond. In 4_2 the spheres are at the vertices of Andreini's polyhedral space filling $3.6^2 + 3.8^2 + 4.6.8$. In 4_3 the spheres are located at the mid-points of edges of the 3_1 net. The sphere packing 4_4 can be obtained by replacing every sphere of the diamond structure 4_1 by a tetrahedron of four spheres.

Koch & Fischer (1995) have systematically identified all possible 3-connected uninodal sphere packings. There are just 52 types (some of which can occur in several

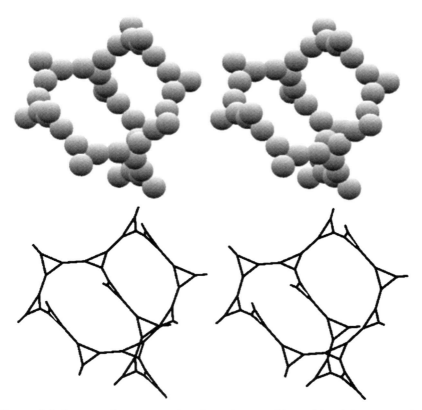

Figure 4.6 Stereo pairs representing the sphere packing with minimum density, and its corresponding 3-connected net.

varieties, with different space group symmetries, on account of the degrees of freedom to move the spheres without disturbing the net topology). The Heesch-Laves configuration 3_2 of lowest density (Figure 4.6) is designated 3/3/c1 in the nomenclature of Koch and Fischer (i.e. connectivity 3; smallest rings triangles; cubic symmetry; the final number 1 distinguishes it from other sphere packings with this description).

4.5 The Boerdijk–Coxeter Helix

Coxeter (1974) suggested a straightforward extension of the concept of a *regular polygon*. A regular polygon as usually defined is a cycle of vertices . . . 1, 2, 3, . . . and edges . . . 12, 23, . . . obtained from a single point by repeated action of a rotation. Coxeter's extension replaces 'rotation' by the more general 'isometry' (distance preserving transformation). A screw transformation generates a *helical polygon* (or *polygonal helix*), an infinite sequence of vertices . . . −1, 0, 1, 2, . . . and edges joining consecutive vertices. A *Coxeter helix* is a polygonal helix such that every set of four consecutive vertices form a regular tetrahedron (Coxeter 1963; 1969). This produces a twisted rod of tetrahedra, the *Boerdijk–Coxeter helix* (Boerdijk 1952; Coxeter 1985). A model can be produced by folding a strip cut from the tiling of the plane by equilateral triangles. Buckminster Fuller (1975) called the helical tower of tetrahedra the *tetrahelix*. Boerdijk (1952) investigated the tetrahelix in connection with dense packings of equal spheres. Boerdijk considered the dense rod-shaped packing in which the sphere centres lie on the vertices of a tetrahelix (Figure 4.7).

As pointed out by Boerdijk, the packing of spheres centred at the vertices of the tetrahelix can be extended by adding more spheres. This determines additional, only slightly irregular, tetrahedra, so that every edge of the Coxeter helical polygon is shared by five tetrahedra. Further extensions of the structure are possible. The

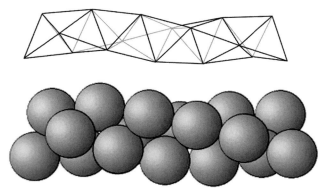

Figure 4.7 The tetrahelix and the corresponding packing of spheres considered by Boerdijk.

next stage of adding spheres gives a rod-like structure in which every sphere of the original structure belongs to a slightly distorted form of the 26-atom γ-brass cluster (Lord & Ranganathan 2001a, b). Another interesting subset of the tetrahedra in this structure is the triplet of distorted B–C helices twisted around each other as shown in Figure 4.8 (Sadoc & Rivier 1999; Lord & Ranganathan 2001a, b).

4.6 Random Sphere Packings

If the assumption of periodicity is dropped, we are faced with the largely unsolved problem of *random* packings of equal spheres, which is of special importance to a proper understanding of the structure of liquids and amorphous solids. It is astonishing that a liquid, such as liquid mercury or water, although apparently disordered, should yet have a definite density. The question as to whether the liquid state is just a disordered crystal structure, or fundamentally different, is of considerable importance.

Reported investigations into *dense random packings* of equal spheres tend to be empirical rather than theoretical – involving either actual experiments with a large collection of hard spheres such as ball bearings, or computer simulations of such experiments. The densities of the resulting arrangement may vary quite widely depending on the method used to produce randomness.

It would appear to be an easy problem to measure the density (the fraction of space filled) of a large mass of equal spheres, poured into a container with curved or irregular walls, and shaken down. If equal ball bearings are poured into a container with flat walls, especially if the walls are set at the correct angles, crystallisation occurs and the usual ccp appears. This has a density of $\pi/\sqrt{18} = 0.74048$. The experimentally observed density of close packed spheres is 0.636 ± 0.001. Less certainly,

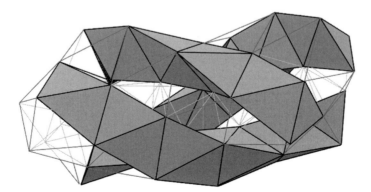

Figure 4.8 Three distorted Boerdijk–Coxeter helices twisted around a central B–C helix. This configuration is the basis for a description of the triple 'coiled coil' structure of collagen (Sadoc & Rivier 1999).

spheres poured into a container, without shaking, have a lower density, about 0.601 (Scott & Kilgour 1969).

Random packings of equal spheres were investigated intensively by Bernal (1959; 1960a, b; 1964a, b). Extensive ball and spoke models were constructed from data obtained from experiments, to facilitate measurement and statistical analysis. Bernal conceived of a monatomic liquid in terms of these packings. In the picture that emerged the atoms tend to cluster in tetrahedral units and aggregates of tetrahedral units (e.g. seven atoms at vertices of a pentagonal bipyramid or eight atoms at the vertices of a 5-tetrahedron portion of a tetrahelix). In Bernal's terminology these are the 'pseudonuclei'. Voids in the structure take the form of deltahedra – convex poly-hedra with equilateral triangular faces, corresponding to voids in the sphere packing in which there is insufficient space for an extra sphere. The voids and the convex pseudonuclei together constitute a set of polyhedra called *deltahedra*. From the Euler formula $F - E + V = 2$ and the condition for triangular faces $3F = 2E$, we have $F = 2(V + 2)$. There are just eight deltahedra (Figure 4.9). The 12-faced deltahedron (Figure 4.9e) is a curious configuration with $\overline{4}\,m$ symmetry. This polyhedron with 12 equilateral triangle faces was first mentioned by Wefelmeier (1937).

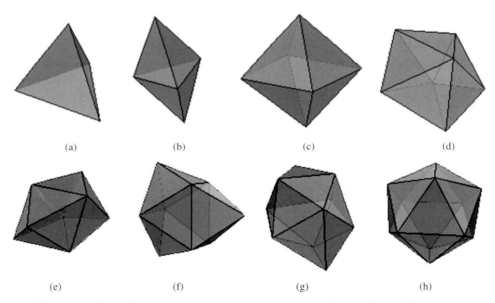

(a)   (b)   (c)   (d)

(e)   (f)   (g)   (h)

Figure 4.9 Bernal's deltaheda: (a) tetrahedron: 4 vertices, 4 faces; (b) trigonal bipyramid: 5 vertices, 6 faces; (c) octahedron: 6 vertices, 8 faces; (d) pentagonal bipyramid: 7 vertices, 10 faces; (e) snub disphenoid (also called the 'Siamese dodec-ahedron'): 8 vertices, 12 faces; (f) triaugmented triangular prism (a triangular prism capped by three square pyramids): 9 vertices, 14 faces; (g) gyroelongated square dipyramid (a square antiprism capped by two square pyramids): 10 vertices, 16 faces; (h) icosahedron: 12 vertices, 20 faces.

Finney, a student of Bernal, has published extensive statistical information derived from two large clusters, one from the work of Scott & Kilgour (1969) and one from the work of Bernal *et al.* (1970). Finney's work gives radial distributions and local density variations for random sphere packings, types of Voronoi regions, the frequencies with which they occur and much more. We mention here only that the density (packing fraction) was found to be about 0.637 and that the 'average Voronoi polyhedron' had 14.25 faces and the average number of vertices per face was 5.16. Finney's concluding remarks emphasise the need for a rigorous *theoretical* approach to random sphere packings.

Gotoh & Finney (1974) have given a convincing calculation of the expected density at 0.6357. They also obtain a lower bound 0.6099 for stable random loose packing. Their argument is that on average each sphere makes six contacts. Choosing a particular direction, contacts with three spheres hold the central sphere from moving in that direction, and three others will prevent motion in the opposite direction. Imagine a box under pressure. The forces on any sphere are in equilibrium. We could resolve the forces on a sphere in three perpendicular directions. Thus we may have six equations for the six forces. The forces between spheres are equal and opposite so that the same forces occur in the relations for the stability for other spheres. If there were only four spheres in contact then the six equations would require that there should be conditions on the angles at which the neighbouring spheres lie. It is probably a coincidence that, within experimental error, the calculated and observed values of the packing density are equal to $2/\pi = 0.6366$ which is the mean value of the cosine.

Torquato *et al.* (2000) have pointed out the lack of rigour in the concept of 'dense random packing' – increase in the density of a packing involves a sacrifice of disorder. They introduced a more well-defined concept, that of a 'maximally random jammed state', which can be made precise with the choice of an order parameter to quantify the randomness.

Concerning *quasiperiodic* packings of equal spheres, worth mentioning are the interesting dense packing discovered by Wills (1990) based on the standard 3D quasilattice of Kowalewski units, and the ingenious hierarchical sphere packings with icosahedral symmetry described by Aston (1999), based on Kramer's (1982) construction.

4.7 Sphere Packings in Higher Dimensions

Packings of equal hyperspheres centred on a lattice in E_n are of great importance in information theory; solutions to the higher dimensional analogues of sphere packing problems provide the foundation for the construction of efficient error-correcting codes (Conway & Sloane 1998). Briefly, when signals comprised of 'words' of length *n* are transmitted over a channel with a poor signal to noise ratio, the words need to be distinguishable from each other in spite of errors. A word can be interpreted as a point in an *n*-dimensional Euclidean space. The problem of designing an effective code is then the problem of minimising the maximum distance between pairs of points in E_n.

Particularly dense structures exist when $n = 8$ and $n = 24$. A projection method applied to the E_8 lattice has been employed in the construction of quasiperiodic structures in E_3 (Sadoc & Mosseri 1993; 1999). *Spherical codes* (Hamkins & Zeger 1997a, b; Conway & Sloane 1998) are obtained from higher dimensional analogues of the Tammes problem to be discussed in the following section.

4.8 Packing Circles on a Spherical Surface

Problems concerning the distribution of subunits over the surface of a sphere are of considerable intrinsic mathematical interest, and are also of importance in certain chemical, biological and engineering contexts. They have given rise to a very extensive literature, only a small selection of which can be cited here.

The Dutch biologist Tammes (1930) studied the distribution of pores on the surfaces of pollen grains, formulating the concept of efficient distribution of the pores as a mathematical problem – the problem of packing *N* congruent non-overlapping circles on a unit sphere, or, *equivalently*, maximising the minimum distance between any pair of *N* points on a sphere. Accordingly, the problem is known as the 'Tammes problem' (Fejes Tóth 1953; 1964).

For certain very special values of *N* the optimum arrangement is known and is highly symmetrical. For $N = 12, 24$ and 60 the points are at the vertices of, respectively, an icosahedron, a snub cube and a snub dodecahedron. Every circle is in contact with five others (the maximum possible number of contacts). Two other highly symmetrical solutions with every circle in contact with five others exist for $N = 48$ and $N = 120$ (Robinson 1961; 1969).

For small values of *N* the distribution of *N* equal-sized atoms around a central atom, forming a 'coordination shell', is an important factor in understanding the formation of metallic glasses (Miracle *et al.* 2003) and poses the same geometrical question.

Mackay *et al.* (1977) presented a compilation of data for the densest known packings of *N* spheres on a spherical surface, complete up to $N = 27$ together with a few higher values. The data was collected from the literature and new values were obtained from an iterative algorithm for maximising the minimum angular separation between pairs of the *N* points. More recent collections of data have been compiled by Clare & Kepert (1991) and Kottwitz (1991). An extensive database for *N* up to 130 has been made available on the Internet by Sloane *et al.* (1995).

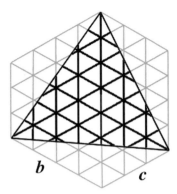

Figure 4.10 A triangular region of a plane hexagonal lattice, of type $b = 4$, $c = 3$.

Variations on the theme of the classic Tammes problem have also been investigated. Applebaum & Weiss (1999) developed a method of approach to the problem of packing identical non-overlapping circles on a *hemisphere*, where the circles are arranged in parallel rings, and obtained optimal solutions up to $N = 40$. They were motivated by the design of the LAser GEOdynamics Satellite (LAGEOS) which carries 426 circular mirrors arranged in this way.

Tarnai & Gáspár (1987) looked for highly symmetrical solutions of the Tammes problem – those with tetrahedral, octahedral or icosahedral symmetry – which occur for special values of N. Their approach was based on an extension of the method of Robinson (1961; 1969). A portion of a plane hexagonal lattice can be inscribed on each triangular face of a tetrahedron, octahedron or icosahedron in the manner indicated in Figure 4.10. If the relationship between a triangular face and the lattice is described by two integers b and c, as demonstrated in the figure for the case $b = 4$, $c = 3$, then the number of vertices per face is $\nu = (b^2 + c^2 + bc)/2$ so that the total number of points on the polyhedron is νF if the vertices of the polyhedron are included and $\nu F - V$ if they are not. The method consists of projecting this point set on to a sphere and then applying an algorithm to maximise the minimum distance between pairs of points. Removing the polyhedron vertices gives denser packings. Tarnai & Gáspár (1987) carried out this program for all three polyhedral symmetries, taking $c = 1, 2, 3$ and $b = c + 1$ or $c + 2$, with and without the polyhedron vertices, and tabulated the packing densities. They found several interesting previously unknown solutions. Two curiosities worth mentioning: for $N = 72$ there are optimal arrangements for each of the three symmetries – tetrahedal, octahedral or icosahedral, and for $N = 54$ the tetrahedral solution and the octahedral solution give the same pattern – which also happens to be a 'fourfold spherical spiral pattern' of the kind first introduced by Székely (1974) that we shall encounter in Section 7.10.

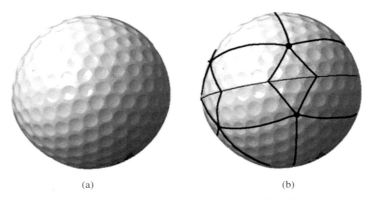

Figure 4.11 (a) Distribution of 332 dimples on a golf ball; (b) this particular arrangement is based on two hemisphers of an icosahedral pattern, related by a reflection.

For aerodynamic reasons the surfaces of golf balls are patterned by a uniform distribution of dimples (Figure 4.11). There is a wide variety of different designs, which may (or may not) have tetrahedral, octahedral or icosahedral symmetry. A special requirement is that one or more great circles must be free of dimples because a golf ball is manufactured from two separate hemispheres. Three orthogonal great circles, for example, partition the surface of a sphere into eight spherical triangular regions and an efficient packing of 32 equal discs produces a golf ball design with 336 dimples and octahedral symmetry. Similarly, a design with 492 dimples is based on icosahedral symmetry and six great circles. Tarnai (1996) has investigated the range of possibilities, their symmetries and numbers of dimples. There is an entertaining short article by Ian Stewart on this topic (Stewart 1997).

The Tammes problem is a member of a wider class of problems concerning the distribution of points on a spherical surface. Consider a collection of points, constrained to lie on a sphere, acted on by mutual repulsion due to a given pair potential. They will assume a stable arrangement for which the total energy is minimised. The literature dealing with this class of problems is very extensive. Particular attention has been focused on pair potentials of the form r^α or $1/\log(r)$ where r is the (Euclidean) distance between two of the points. For the (formidable) mathematical details we refer the reader to the work of Rakhmanov *et al.* (1995) or Kuijlaars & Saff (1998) and works cited by them. The optimal arrangement of the N points will depend, in general, on the particular choice of potential. For $\alpha = -1$ we have the classic case of the Coulomb potential; the question is then how N electrons constrained to lie on a sphere will arrange themselves under the effect of their mutual electrostatic repulsion. Rakhmanov, Saff and Zhou (1994) have carried out extensive numerical experiments for the cases $\alpha = -2, -1, 1, 2$ and the logarithmic case

('$\alpha = 0$') up to $N = 200$ and tabulated the resulting extremal energies and the symmetries of the configurations. When the exponent α takes large negative values the forces between nearest neighbours dominate in the expression for total energy and minimising the total energy then becomes equivalent to maximising the smallest distance. In other words, the Tammes problem is an asymptotic limiting case of these more general problems (Melnyk *et al.* 1977).

Bausch *et al.* (2003) have experimentally investigated the minimal energy arrangements of packed equal spheres in contact with a spherical surface. They studied the patterns assumed by small (1 μm) polystyrene beads adsorbed on the surface of a water droplet (suspended in a density-matched toluene–chlorobenzine mixture). When N is large most of the surface shows a hexagonally close packed arrangement, which necessarily has defects. For topological reasons there must at least be 12 spheres with fivefold coordination, and as N increases more defects, positive disclinations (coordination 5) and negative disclinations (coordination 7), will occur to enable the hexagonal array to fit the curved surface (recall the discussion of Sections 2.11 and 2.12). Theory predicts that the number of excess defects depends linearly on N. An intriguing aspect of this investigation is the discovery that the defects occur in chains (-5-7-5-7-) forming *grain boundaries* between areas of hexagonal arrays as in Figure 4.12 (unlike grain boundaries in the case of *flat* 2D cystals, these 'scars' can have end points).

Some of the reasons for the difficulty and the fascination of these problems lies in the fact that there are, in general, multiple solutions. There may, for a given potential and a given N, be many stable configurations (i.e. configurations for which the energy is at a minimum but not the absolute minimum). One cannot be certain whether a plausible solution is actually the 'best' solution for the given N. A fascinating instance of this has been described by Fowler & Tarnai (1999); the problem

Figure 4.12 A portion of a 'spherical crystal' showing a grain boundary consisting of a chain of positive and negative disclinations.

dealt with here concerns the *minimising* of the radius of *N overlapping* circles that
cover the surface of a sphere and the transition from a packing to a covering, as the
radius is increased. The work deals specifically with $N = 13$ and takes into consideration no fewer than 28 different optimal arrangements.

In the Tammes problem the number N is given and an optimum radius is to be
found. Conversely, we may be given a radius, and ask for the number N. The question then is: what is the greatest number of spheres of unit radius that can be in contact with a central sphere of given radius R. The case $R = 1$ has an interesting history.
Twelve equal spheres can be placed in contact with a central sphere of the same
radius. A discussion took place in 1694 between Isaac Newton and David Gregory.
Gregory maintained that an arrangement of the 12 spheres around the central sphere
might be possible, that would leave room for a 13th sphere to be fitted in. Newton
conjectured that it isn't possible, but the conjecture remained unproven until the
nineteenth century. A recent proof was published by Leech (1956) and Conway &
Sloane (1988). The problem is non-trivial because the 12 spheres are free to move
around without losing contact with the central sphere – there are an infinite number
of configurations. In particular, there is a curious transformation of the cuboctahedral
configuration that converts it to an icosahedral arrangement by a cooperative motion
of all 12 spheres – involving rotation of the eight triangular faces (Figure 4.13).
Applied to the 13 sphere cluster the transformation also moves the 12 spheres apart
from each other to keep the central sphere radius unchanged. The transformation is a
spherical analogue of the collective motion of the packing of circles in a plane that
was illustrated in Figure 4.3. Buckminster Fuller's 'jitterbug' (Buckminster Fuller
1975) illustrates the transformation by a construction of 24 rigid rods connected by
flexible joints. (An interesting variant relates the dodecahedron and the polyhedron
5.6^2 (Kovacs & Tarnai 2001).) A third interpretation of this relation between the

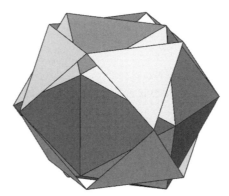

Figure 4.13 Buckminster Fuller's 'jitterbug' transformation relating the cuboctahedron and the icosahedron.

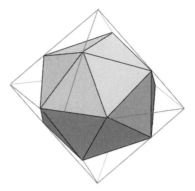

Figure 4.14 An icosahedron inscribed in an octahedron.

cuboctahedron and the icosahedron comes from considering the division of the edges of an octahedron in the ratio $\rho : 1$. The cuboctahedron and icosahedron correspond, respectively, to $\rho = 1$ and $\rho = \tau = (1 + \sqrt{5})/2$ (Figure 4.14).

4.9 Packing of Unequal Spheres

Whereas the problems arising from the packing of *equal* spheres have a long hisory and have generated very extensive mathematical literature, much less is known about the properties of packings of unequal spheres. An adequate theory would be of enormous value to materials science. The atoms and ions of particular elements can be assigned characteristic radii (Pauling 1960). Though the 'radius of an atom' is not a sharply defined quantity – real atoms are nothing like 'hard spheres' – the concept can account for limitations on the possible distances between atoms, provide estimates of bond lengths, and useful insights into the structure of a mineral or a glass can be obtained from considering homologous packings of hard spheres.

The two dimensional analogues – packings of unequal circles on the Euclidean plane – occur naturally, for example in granular materials, in turbulent liquids, in vapour depositions, etc. The size distribution of circles of radius r follows a power law distribution $N(r) = r^{-\alpha}$. Aste has developed a theoretical approach to the topological and geometrical rules underlying these packings. He has shown, for example, that the maximum value of α is 2 (and for D-dimensional sphere packings, it is D), and obtained an approximate formula relating α to the average coordination number of the circles (Aste 1996).

Examples of packings of spheres that are not all the same size come from considering the sizes of the interstices in packings of equal spheres. Some simple examples are illustrated in Plate VI. Consider for example the ccp arrangement of spheres of unit radius. This is the densest possible packing of *equal* spheres, with a packing

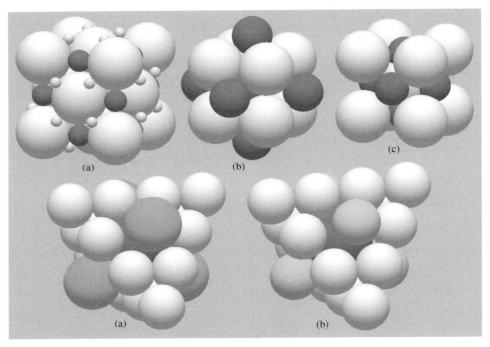

Plate VI Top row: (a) the fcc sphere packing with smaller spheres in the octahedral interstices (red) and in the tetrahedral interstices (green), (b) spheres occupying the voids in a primitive cubic array of close packed spheres, (c) a bcc array of spheres with smaller spheres occupying the voids. Bottom row: (a) a portion of the Laves type packing of spheres of three kinds, with optimised packing fraction, (b) the modified AB_5 Laves phase in which half the large atoms are replaced by the small atoms.

fraction $\pi/\sqrt{18} \sim 0.7405$. The sphere centres lie at the vertices of an fcc lattice (i.e. at the vertices of a regular space filling of octahedra and tetrahedra). Thus smaller spheres can be inserted in the octahedral voids, up to a maximum radius $\sqrt{2} - 1 \sim 0.414$. The packing fraction is then increased to 0.816. The tetrahedral voids can accommodate still smaller spheres, up to a radius $\sqrt{(3/2)} - 1 \sim 0.225$, and the packing fraction is increased a little more, to 0.825. Similarly, a close packing of unit spheres centred on a primitive cubic lattice has packing fraction $\pi/6 \sim 0.524$; the interstices can accommodate spheres of radius $\sqrt{3} - 1 \sim 0.732$, increasing the packing fraction to 0.729. In a close packed array of unit spheres centred on a body-centred cubic (bcc) lattice the interstices can accommodate spheres of radius $1/\sqrt{3}$, increasing the packing fraction from 0.608 to 0.702.

A more interesting example of a regular close packing of equal spheres with smaller spheres occupying the voids is the hard sphere model of the cubic Laves phase. Recall that the large atoms in a cubic Laves phase are located at the centres of the truncated tetrahedra and the smaller atoms at the vertices in the space filling $Fd\bar{3}m$: $3^3 + 3.6^2$ (Figures 3.5 and 3.17). The smaller atoms, at the vertices, can be represented by spheres each touching six others. The large voids can then be occupied by larger spheres, each touching four other large spheres. The ratio of radii is then $\sqrt{(3/2)} \sim 1.2247$ and the packing fraction is 0.710. This is, then, an 'ideal' ratio of atomic sizes, for the formation of a Laves phase. In actuality there can be considerable deviations from this value because, of course, real atoms are not 'hard spheres'. A variant of this arrangement, with three different sphere sizes, is possible in which the large spheres are of two different sizes, in an alternating arrangement (*cf.* the structure of zinc blende ZnS compared to that of diamond). The greatest packing fraction is obtained when 50% of the large spheres touch the small spheres. The radius ratios are then $\sqrt{11} - \sqrt{2} : \sqrt{12} - \sqrt{11} + \sqrt{2} : \sqrt{2} \sim 1.345 : 1.104 : 1$, with packing fraction 0.8846. Laves phase with three atomic sizes have been produced, which approximate to this 'ideal' arrangement. Another variant of the cubic Laves phase of type AB_5 occurs when 50% of the large atoms (atoms A in AB_2) are replaced by small atoms. Examples are $ZrCu_5$ and $AuBe_5$.

Hudson (1949) considered the interesting problem that comes from considering how smaller spheres can be packed into the interstices of a close packing of equal spheres. Close-packed arrangements of equal spheres have two kinds of voids – those surrounded by a tetrahedral arrangement of four spheres and those surrounded by an octahedral arrangement of six spheres. For a close packing of unit spheres there is, for any given number N, a maximum radius r such that N spheres of radius r can be packed in each octahedral void. Hudson derived a value of r for N up to 27 and found an intricate sequence of optimal arrangements and associated maximum packing fractions.

It has become apparent in recent years that the relative atomic sizes and the proportions of the various elements in an alloy have a major role in determining the kind of clustering that will take place in a complex alloy, with a consequent influence on the possibility of forming glassy (amorphous) or quasicrystalline structures (see Senkov & Miracle 2001; Miracle *et al.* 2003). Briefly, the pattern that emerges from this work is that in ternary alloys, glass-forming ability is enhanced if the smallest and largest atoms are more abundant (plot of atomic size versus atomic percentage is concave upward) and hindered if the middle-sized atom is the most abundant (concave downwards). Miracle (2004) has developed a structural model for metallic glasses, based on the dense packing of atomic clusters. The solute atoms are surrounded by a coordination shell of solvent atoms, forming a cluster. The coordination number is of course governed by the relative sizes of the atoms involved. The proposal is that, in a metallic glass these clusters arrange themselves in a regular manner (typically, fcc or hexagonal close packed) and that the randomness characteristic of an amorphous material is a feature of the arrangement of the solvent atoms. The model is supported by considerable experimental evidence.

4.10 Rod Packings

The possibilities for dense packings of infinite cylindrical rods have been studied by various investigators. A few examples are shown in Figure 4.15. Particularly interesting are those with cubic space group symmetries (O'Keeffe 1992; O'Keeffe *et al.* 2001; 2002). Remarkable quasiperiodic rod packings with icosahedral symmetry have been found by Duneau & Audier (1999) and Audier & Duneau (2000).

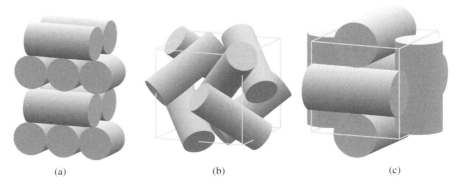

(a)   (b)   (c)

Figure 4.15 (a) A portion of a simple rod packing; its space group symmetry is $P4_2/mmc$, (b) a unit cell of a rod packing with symmetry $Ia\overline{3}d$, the rods run along threefold axes; this is the densest of all the rod packings, with a density $\rho = (\pi\sqrt{3})/8 = 0.68 \ldots$ (c) a unit cell of a rod packing with symmetry $Im\overline{3}$.

5

Hierarchical Structures

In Chapter 2 we mentioned a pattern of regular pentagons (Figure 2.10) generated by iteration of a simple rule whereby each pentagon is converted to a cluster of six pentagons. If the single initial pentagon is called the 'zeroth-order' pattern, then the Nth-order pattern is a symmetrical arrangement of 6^N pentagons. The pattern is *hierarchical* in the sense that at each stage the structure is built from subunits that are replicas of the pattern belonging to the earlier stages.

There is, clearly, an infinite variety of possibilities for generating patterns and structures from an iterative rule or set of rules applied to one or more primitive shapes. Some striking examples of the emergence of complexly organised geometrical structures from the iteration of simple local rules are the 'shape grammars' of Gips (1975) and Stiny (1975) and Stanislav Ulam's modular patterns (Ulam 1966; Schrandt & Ulam 1970). Hierarchical structures or fractal structures result from iterative procedures that involve rescaling (Whyte *et al.* 1969; Mandelbrot 1982; Peitgen & Richter 1986; Sander 1987; Prusinkiewicz & Lindenmayer 1990). We shall indicate a few particularly interesting examples – some of which are relevant to the understanding of shapes and patterns in nature – from quasicrystals to biological forms.

5.1 Lindenmayer Systems

The fractal curves we shall describe and illustrate in this chapter are particular example of structures derivable from *Lindenmayer systems*, or *L-systems*, introduced by Lindenmayer (1968). Lindenmayer identified and systematised the underlying formal logic of iterative computer graphics. At an abstract level an L-system consists a set of symbols ('letters'), strings of symbols ('words') and iterative dynamical rules for replacing letters by words. A very simple example was described in Section 2.8, an L-system with two letters A and B and the dynamical rules $B \rightarrow A$, $A \rightarrow AB$, leading to the sequence of symbol strings B, A, AB, ABA,

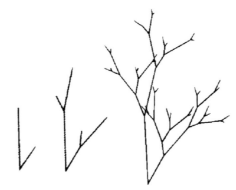

Figure 5.1 A simple tree-like graphic image produced by iteration (zeroth, first and fourth order).

ABAAB, …. The formal language of an L-system is then supplemented by inter-preting the letters as commands in a graphic system, to create an image. The FRACTINT software, available on the Internet, employs 'turtle graphics', convert-ing a string of commands into a fractal image.

Prusinkiewicz and Lindenmayer have developed this approach into a highly sophisticated means of simulating plant morphology. Their classic work is pre-sented in *The Algorithmic Beauty of Plants* (Prusinkiewicz & Lindenmayer 1990). Surprisingly realistic images resembling the branching structures of plant forms can be obtained from very simple iterative procedures, involving a rescaling at each iter-ation (Figure 5.1). Further elaboration of the basic method can eliminate the artifi-cial look produced by the strict self-similarity of the resulting images. Stochastic L-systems employ random choices of iterative rule, with assigned probabilities. The effect of interaction of plants with their environment can be simulated (Mech & Prusinkiewicz 1996). The two-dimensional (2D) graphics can be generalised to 3D-form generation. L-systems are developing into a valuable tool for investigating morphogenetic development in living systems.

These developments lie outside the aims and scope of the present work. We shall therefore focus attention only on a few interesting patterns that arise when a single simple transformation is repeatedly applied to a simple starting image, lead-ing in the limit to a fractal with a strictly hierarchical structure.

5.2 Fractal Curves

Hilbert's curve has the curious property of being a 1D object (a plane curve) that nevertheless passes at some stage arbitrarily close to any chosen point in the interior of a square. A simple way of defining it is to construct it from an iterative procedure

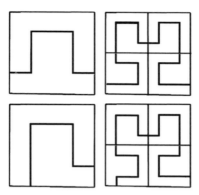

Figure 5.2 The iterative procedure that produces the Hilbert curve from the decoration of a tiling pattern.

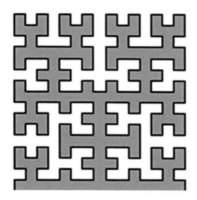

Figure 5.3 The Hilbert curve after three iterations. The two sides of the curve are differently coloured for clarity.

applied to a pair of square tiles decorated by lines. The two tiles are shown on the left in Figure 5.2. The iteration converts each of these tiles to the group of four tiles shown on its right. The results after three iterations and after five iterations are shown, respectively, in Figures 5.3 and 5.4. The pathological Hilbert curve with the 'square-filling' property is of course the unimaginable limit of this process.

The *dragon curve* can, like the Hilbert curve, be generated by an iterative procedure. The primitive unit (zeroth-order dragon curve) is simply a line segment. The Nth-order curve is obtained consists of two curves of order $(N - 1)$ joined end to end and related to each other by a 90° rotation. It follows that the lower-order dragon curves can be made by repeatedly folding a long strip of paper in half, left-hand end to right-hand end and opening out the result so that all the folds make right angles. (Allegedly, this is how the curve was serendipitously discovered.) The

Figure 5.4   The fifth-order Hilbert curve.

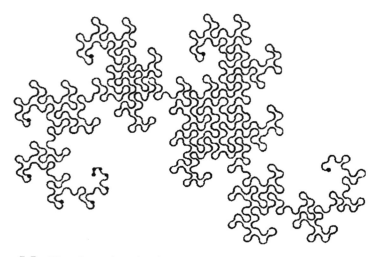

Figure 5.5   The eleventh-order dragon curve.

result is indicated in Figure 5.5. The right-angled corners have been rounded to indicate the continuity of the curve – with sharp corners the curve touches itself at numerous points. The blobs indicate the end points of the curves of order 0, 1, 2, 3, etc. They lie on a logarithmic spiral.

   The iterative rule shown in Figure 5.6 converts a line segment to a pair of line segments (the grey triangles are 'markers' to indicate orientation). The dragon curves arises from application of this transformation to every line segment, at each iteration.

Figure 5.6 The rule for generating a dragon curve from a line segment.

Figure 5.7 The sixteenth-order dragon curve. The mesh of squares that result from the self-contacts at numerous corners have been filled in to emphasise the shape of the boundary. The limiting fractal shape can tile the plane.

This produces a sequence of curves all having the same two-end points, their length increasing by a factor $\sqrt{2}$ at each iteration. In the limit, the dragon curve is a fractal object which will tile the plane. This is indicated in Figure 5.7 and Plate VII.

The curve can be encoded neatly as a sequence of 'turn left' and 'turn right' instructions. If s is a string of such instructions (denoted by L and R) and if s' denotes the sequence obtained by reversing the order and interchanging L and R, then the dragon curves of increasing order are given by the algorithm $s_0 = R$, $s_{n+1} = s_n R s_n'$.

Perhaps the most well-known fractal curve is *Koch's snowflake curve* (Figure 5.8), which in the limit has infinite length but encloses a finite area. It is generated simply by starting with an equilateral triangle and at each iteration placing smaller equilateral triangles on every edge of the growing configuration. Figure 5.9 indicates the iterative rule that gives rise to *Peano's curve* (Peano 1890), which has the

Plate VII A tile produced from four dragon curves.

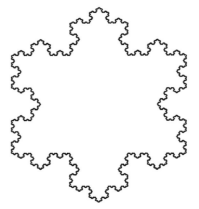

Figure 5.8 The snowflake curve.

same 'square-filling property' as Hilbert's curve. Finally, Figure 5.10 shows the iterative rule for the '*flowsnake*', Figure 5.11. Observe how these curves can arise in the decorations of a hexagonal tiling patterns in which the tiles are all identically decorated.

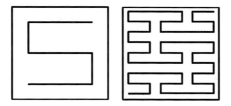

Figure 5.9 The generation of Peano's curve.

Figure 5.10 Iterative rule for the flowsnake.

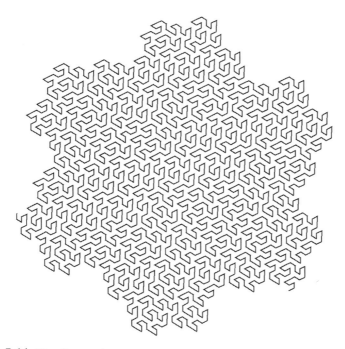

Figure 5.11 The flowsnake.

5.3 Inflation Rules

The Penrose tiling patterns can be generated in several ways: by imposing 'matching rules' on the tile edges so that periodicity cannot arise when assembling the pattern, they can be obtained by projection of a slice of the 5D-hypercubic lattice onto a plane, or they can be generated by an *inflation*. In the inflation method for the rhomb tilings, the two-rhombic tiles are partitioned into patches of a Penrose pattern on a smaller scale (smaller by a factor $\tau^{-1} = (\sqrt{5} - 1)/2$), in the manner indicated in Figure 5.12. Beginning from a single tile of either kind, and *inflating* the pattern by a factor $\tau = (1 + \sqrt{5})/2$ after each iteration, produces a growing patch of the aperiodic tiling pattern that in the limit extends over the whole plane.

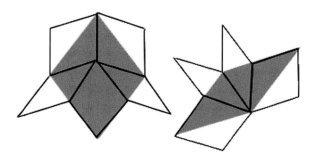

Figure 5.12 The inflation rule for the Penrose rhomb tilings.

5.4 Other Aperiodic Tilings by Iteration

The edges of a Penrose rhomb pattern have only five different orientations. The patterns are said to have a long-range order with fivefold symmetry. The icosahedral and decagonal quasicrystals also have long-range orientational order with five or tenfold symmetry. Fivefold symmetry is not possible for periodic structures, so that in the early days after the discovery of quasicrystals, the appearance of 'forbidden' symmetry in their diffraction patterns was thought to be a defining characteristic of these materials. This is not so – many kinds of quasicrystal with orientational order of a kind allowed by conventional crystallography are now known. In this section we briefly mention a few aperiodic tiling patterns of other than Penrose type, that can be obtained by iteration.

In Figure 5.13 we demonstrate an inflation rule applied to only a *single* kind of tile (a 60° rhombus) that produces an aperiodic tiling with a hexagonal symmetry. The result after four iterations is shown in Figure 5.14.

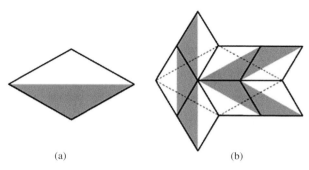

(a)   (b)

Figure 5.13 (a) The fundamental tile for an aperiodic pattern with sixfold symmetry, (b) the result of the first iteration.

Figure 5.14 The hexagonal pattern after four iterations.

5.5 Inflation of 3D-Penrose Tilings

The 3D Penrose tilings, or 'Ammann tilings', constructed from the two Kowalewski rhombohedra, were introduced in Section 3.10. The standard method for generation of these tilings is by a projection onto three space from a slice of a 6D hypercubic lattice. The tilings have a τ^3 inflation rule (Ogawa 1986; Audier & Guyot 1988); that is, any such tiling with tiles of unit edge length can be decorated by superimposing on it a

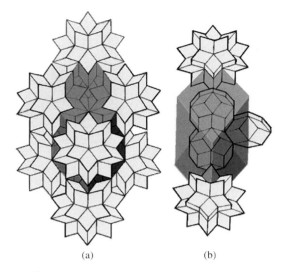

Figure 5.15 The τ^3 inflation rule for the prolate unit (shown dark grey transparent) of a 3D-Penrose pattern. (a) A 'star polyhedron' built from 20 of the small prolate units is centred at *every vertex* of the larger scale pattern, (b) a *rhombic icosahedron* is centred at the mid-point of each edge the larger scale pattern. The decoration of the large prolate unit is completed by a pair of *rhombic triacontahedra* sharing a single oblate unit, with centres on the major diagonal.

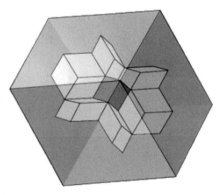

Figure 5.16 The τ^3 inflation rule for the oblate unit: a small prolate unit and ring of six small oblate units inside each oblate tile.

similar tiling with tiles of edge length τ^{-3}, in such a way that all the prolate units are decorated (subdivided) in an identical way by the smaller tiles, as are all the oblate units. Thus, as for the 2D-Penrose tilings, the aperiodic patterns can be produced iteratively by repeatedly subdividing and inflating by a factor τ^3. An insight into some of the details of the relation between the unit of edge-length 1 and the related edge-length τ^3 pattern is a valuable aid to the understanding of icosahedral quasicrystals, as Audier and Guyot showed – see also Lord *et al.* (2000). Figures 5.15 and 5.16 explain it. A *star polyhedron* built from 20 of the small prolate units is centred at *every vertex* of the larger scale pattern and a *rhombic icosahedron* is centred at the mid-point of every edge of the larger scale pattern. This prescription leaves a void inside the large prolate units into which fit a pair of triacontahedra sharing a common oblate unit. This is illustrated in Figure 5.15. For a large oblate unit, the star polyhedra centred at the two ends of the threefold axis have a small prolate unit in common (darkly shaded in Figure 5.16). The star polyhedra at vertices and rhombic icosahedra on edges then fill the large oblate unit except for a ring of six small oblate units shown in Figure 5.16.

5.6 Kramer's Hierarchical Tiling

An interesting geometrical curiosity discovered by Peter Kramer (1982) is of considerable importance in principle in that it illustrates hierarchical organisation as a way, more general than periodic repetition, of building infinite structures from a small number of components. Note that the outer vertices of a stellated icosahedron form a dodecahedron, and vice versa (Kepler 1619; Wenninger 1971; Cundy & Rollett 1951, 1961) (Figure 5.17). Starting from a central dodecahedral tile and repeatedly stellating, one can obtain a subdivision of space with a hierarchical structure, consisting of seven different tile shapes. Kramer showed how each of these seven shapes can be built up from smaller replicas of the same seven shapes, and hence deduced a tiling of E_3, with icosahedral symmetry, built from seven kinds of tile.

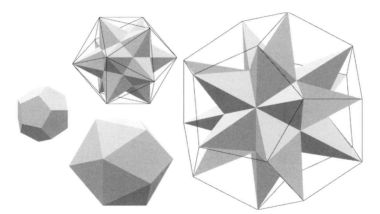

Figure 5.17 The successive stellation of a dodecahedron, giving a dodecahedron larger by a factor τ^3, where $\tau = (1 + \sqrt{5})/2$. (The stellated polyhedra are Kepler's $\{5/2, 5\}$ and $\{5/2, 3\}$.)

6

Clusters

Various crystalline phases of alloys with large unit cells containing very many atoms have been elucidated in terms of atomic clusters with icosahedral symmetry. These phases are structurally closely related to the quasicrystals that tend to accompany them. $MoAl_{12}$ (Figure 6.1) is a simple example of icosahedral clustering in a body-centered cubic (bcc) structure (Pauling 1960). Simple 12-atom icosahedral clusters combine in various ways in the several phases exhibited by boron (Pauling 1960; Sullinger & Kennard 1966). The simplest is the rhombohedral form shown in Figure 6.2.

6.1 Clusters of Icosahedra

A large number of highly complex intermetallic phases with hexagonal symmetries have been described in terms of a few basic units, formed from 12- or 13-atom icosahedral clusters which share atoms and may interpenetrate (Kreiner & Franzen 1995;

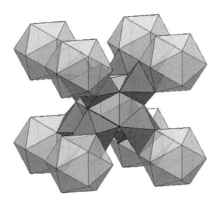

Figure 6.1 The structure of $MoAl_{12}$. Aluminium icosahedra surrounding molybdenum atoms, the icosahedral clusters linked along [111] directions by octahedral linkages.

Kreiner & Schäpers 1997). A basic unit is the *i*3 cluster, Figure 6.3. Note that the threefold axis of this structure lies along twofold axes of the constituent icosahedra; more surprising is the near-coincidence of fivefold and twofold icosahedral axes, perpendicular to the threefold axis of the unit. Other units identified as building blocks in complex intermetallic phases are shown in Figure 6.4.

The curious tetrahedral structure in Figure 6.5, comprised of six double icosahedra linked by sharing atoms, occurs in the formidable structure of Cd_3Cu_4 elucidated by Samson (1967a). Three more atoms over each face of this unit will complete four additional icosahedra. See Kreiner and Franzen (1995), or Samson's original paper, for further details of this remarkable crystal structure.

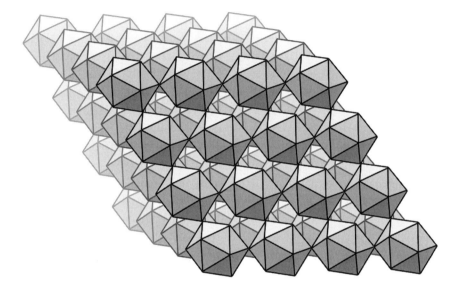

Figure 6.2 The structure of α-boron. The icosahedral clusters are linked by sharing atoms, to form rhombohedral crystals.

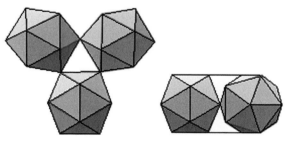

Figure 6.3 The *i*3 unit, viewed along the threefold axis and perpendicular to the threefold axis.

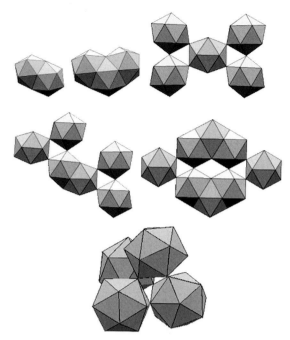

Figure 6.4 Building blocks in complex intermetallic phases, identified by Kreiner and Franzen. The first picture shows the 19-atom 'double icosahedron'. The final picture shows the remarkable tetrahedral '*L*-unit'. Observe that the interior void is octahedral and that half-octahedra, with only slight deformation, fit into the three faces of the unit.

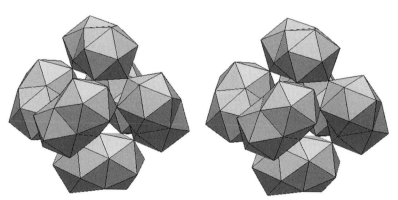

Figure 6.5 Stereo images of a tetrahedral cluster formed from six double icosahedra.

Silver and gold atoms can form 13-atom icosahedral clusters. Teo & Zhang (1991) have described a plausible growth mechanism for successively larger 'clusters of clusters', leading to the '*i*13' supercluster, Figure 6.6. For clarity, the thirteenth sphere, in the centre, is omitted from the figure. The 32 innermost atoms of this structure form a

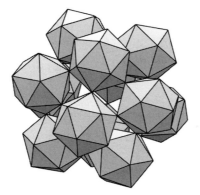

Figure 6.6 A cluster of 12 vertex-sharing icosahedra.

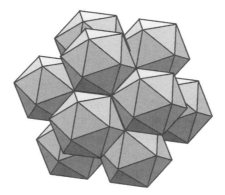

Figure 6.7 A cluster of 12 edge-sharing icosahedra.

Mackay cluster. Twelve *edge-sharing* icosahedra form an interesting geometrical con-figuration (Figure 6.7), an icosahedron-shaped cluster with an edge length $\tau^2 = \tau + 1$ times the edge length of the component icosahedra. The 12 inner vertices form a smaller icosahedron of edge length $\tau^{-1} = \tau - 1$ (τ is the 'golden number $(1 + \sqrt{5})/2$). This kind of cluster was described by Hiraga *et al.* (1985) in connection with their suggested model for icosahedral Al–Mn ('Schechtmanite'). An iteration of the relation between the large outer icosahedron and the 12 icosahedra contained in it gives rise to a hierarchical pattern of icosahedra, somewhat analogous to the two dimensional (2D) pattern of pentagons that we illustrated in Figure 2.11.

6.2 The Bergman Cluster

An icosahedral cluster of 12 spheres can accommodate a second, dodecahedral, shell of 20 spheres each lying over a face of the inner icosahedron and making contact

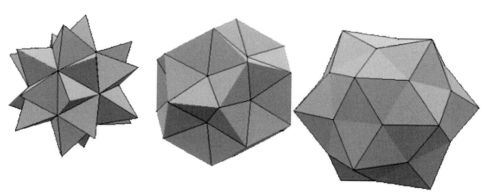

Figure 6.8 A cluster of 130 tetrahedra packed around an icosahedron: (a) a regular tetrahedron on each face of an icosahedron; (b) additional tetrahedra completing the 'five-rings' (*cf* Figure 3.6) and (c) a five-ring fitted into each of the concave faces of the dodecahedron in (b).

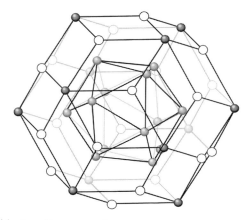

Figure 6.9 The 44-atom Bergman cluster.

with three of its spheres. If a third shell of 12 spheres is added, by placing spheres in the faces of this dodecahedral cluster, in contact with the spheres of the inner icosahedron, their centres form an icosahedron approximately twice the size of the icosahedron of the first shell. The result is a polytetrahedral cluster of 44 equal spheres, represented in Figure 6.8 by the corresponding packing of tetrahedra whose vertices are the centres of the spheres. The spheres of the second, dodecahedral, shell can be some 20% larger, to bring them into contact with the three spheres of the outer icosahedron. The cluster shape then closely approximates to a rhombic triacontahedron. The cluster occurs as an arrangement of atoms in a variety of metallic alloys. It is known as the *Bergman cluster* or *Pauling triacontahedron* (Bergman *et al.* 1952, 1957) (Figure 6.9).

Figure 6.10 The Samson cluster; 104 atoms. A packing of 20 truncated tetrahedra. Placing atoms at the vertices and centres of the truncated tetrahedra gives a Bergman cluster surrounded by a truncated icosahedron.

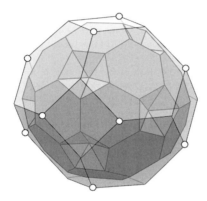

Figure 6.11 A truncated octahedron 4.6^2 with an inscribed 5.6^2.

The Bergman cluster can be surrounded by 60 more atoms, two over each face of the triacontahedron, giving a shell in the form of a truncated icosahedron (5.6^2) (i.e. a 'Bucky-ball' structure). Up to this stage, the cluster is the Samson complex (Samson 1972) (Figure 6.10). If a (5.6^2) is only slightly deformed, Figure 6.11 shows how all its vertices can lie on the faces of a truncated octahedron (4.6^2). Just 12 of the 24 vertices of this (4.6^2) are occupied by atoms, as shown in the figure. Since the truncated octahedron is a 'space-filling' polyhedron, we see how these 117-atom clusters can fit together, sharing their 72 outer atoms, to form a bcc array. Sixty of the outer atoms are shared between two clusters, the other 12 are shared by four clusters. Since the bcc lattice has two lattice points per unit cell, it follows that the number of atoms per unit cell in this idealised structure is $2 \times (44 + \frac{1}{2}60 + \frac{1}{4}12) = 154$. This

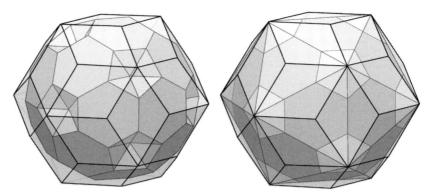

Figure 6.12 The geometry of the R-phase: a Samson cluster in a triacontahedron. Observe that the fivefold triacontahedral vertices complete the rings of five tetrahedra, so that the overall shape of the Samson unit is extended to an icosahedron.

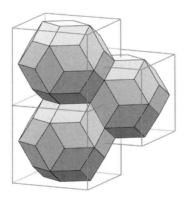

Figure 6.13 The geometry of the R-phase. Interpenetrating triacontahedra in a bcc array, in face contact in the [100] directions and sharing an 'oblate Kowalewski unit' in [111] directions – the inner truncated icosahedron shown in Figure 6.12 is slightly distorted so that 12 of its vertices lie in pairs on the faces of the cube and eight of the hexagonal faces lie inside oblate Kowalewski units. The inner and outer triacontahedra have common vertices – see Plate VII.

structure, known as the R-phase, was first identified by Bergman *et al.* (1952; 1957); see also Audier *et al.* (1998).

    In a metrically slightly different, but *topologically* equivalent model of an R-phase (Komura *et al.* 1960; Audier *et al.* 1988; Romeu & Aragon 1993) the 104-atom clusters occur inside a *triacontahedral shell*, as in Figure 6.12, $\tau = (1 + \sqrt{5})/2$ times larger than the inner Pauling triacontahedron. These 136-atom clusters are centred on

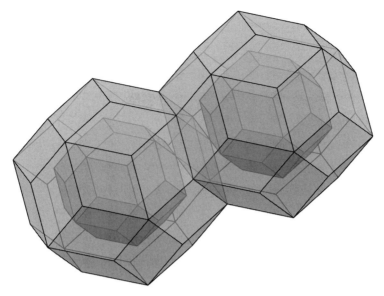

Plate VIII Two of the triacontahedral clusters of the R-phase, showing the 'twinning' along a threefold axis. Note how the inner and outer triacontahedra have common vertices.

a bcc lattice (Figure 6.13). They interpenetrate: eight of the vertices of the small triacontahedra are also vertices of large triacontahedra (Plate VIII).

A description of a complex crystalline solid in terms of clusters of atoms in the form of a sequence of *nested polyhedral shells* (Chabot *et al.* 1981) appears to have been first adopted in the description of γ-brass given by Bradley & Jones (1933). Since then, this kind of description has become of central importance to our understanding of the way in which intricate crystalline structures with large unit cells can arise. Hiraga *et al.* (1998; 1999) have described a variety of cubic intermetallic phases with large unit cells that are built from large clusters centred around Bergman clusters.

6.3 The Mackay Icosahedron

As we have seen, the 44-atom Bergman cluster can be arrived at by packing tetrahedra around an icosahedron. The *Mackay icosahedron* is obtained by packing tetrahedra and octahedra around an icosahedron. Specifically, spheres can be packed around an icosahedral 12-sphere cluster by adding layers, over every icosahedron face, in a configuration that is very nearly cubic close packing (Mackay 1962). The dihedral angles of the icosahedron and the octahedron are $\theta_{\text{icos}} = 138.19°$ and $\theta_{\text{oct}} = 109.47°$.

Hence $360° - (\theta_{icos} + 2\theta_{oct}) = 2.87°$. Thus, if an octahedron is placed on every face of an icosahedron, the angular gap between neighbouring octahedra can be closed by a very small deformation, to bring them into face contact (Figure 6.14). The concave regions of the resulting polyhedron can be filled by five-rings of tetrahedra (Figure 6.15). There are now 54 vertices, internal and external, representing a 54-atom cluster of spheres. The process can be continued indefinitely. The number of spheres in the nth layer is $10n^2 + 2$. The total number of spheres in the cluster, including the nth layer, is $n(10n^2 + 15n + 11)/3$. The 54-atom Mackay cluster, Figure 6.16, is a common structural feature in complex alloys, including quasicrystals.

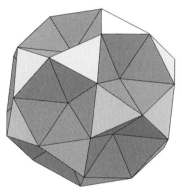

Figure 6.14 Twenty octahedra on the faces of an icosahedron.

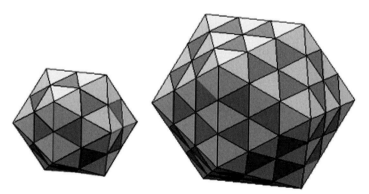

Figure 6.15 The geometry of the Mackay icosahedron: (a) 20 octahedra around an icosahedron, giving an icosidodecahedron; (b) 60 tetrahedra filling the concavities of the configuration Figure 6.14; (c) layers of octahedra and tetrahedra can be added indefinitely.

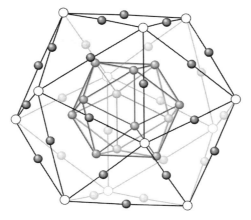

Figure 6.16 The 54-atom Mackay cluster.

6.4 The γ-Brass Cluster

Bradley and Thewlis (1926) identified the structure of γ-brass and described it in terms of a $3 \times 3 \times 3$ cubic cell built from 27 cubic unit cells of a bcc lattice. The lattice points at the vertices of this $3 \times 3 \times 3$ block and at its centre are vacant. We get a structure with a unit cell containing $27 \times 2 - 2 = 52$ sites. The positions of the 52 atoms in a unit cell of γ-brass were understood in terms of their deviations from these 52 sites. In *The Nature of the Chemical Bond* Pauling (1960) describes the $3 \times 3 \times 3$ model of γ-brass. He then casually makes the cryptic remark: 'the structure is an icosahedral one'.

The structure is actually polytetrahedral. The 52 positions in the unit cell are accounted for in terms of 26-atom clusters centred on a bcc lattice. Nyman & Andersson (1979) described the 26-atom cluster as a packing of equal spheres (a convenient idealisation – in the γ-alloys, of course, the spheres (atoms) are of more than one kind). Whereas the Bergman cluster is a polytetrahedral structure systematically built around a single vertex, the γ-brass cluster starts from a single tetrahedron.

Place four spheres in contact. Then place a sphere over each face of the tetrahedral cluster. The centres and bonds then form a '*stella quadrangula*' built from five regular tetrahedra. Six more spheres placed over the edges of the original tetrahedron form an octahedral shell. In terms of the network of centres and bonds we now have added 12 more tetrahedra, which are not quite regular. There are now five tetrahedra around each edge of the original tetrahedron (Figure 6.17). Finally, 12 more spheres complete the rings of five tetrahedra around the edges of the four secondary tetrahedra (Figure 6.18). The 26-atom cluster is more commonly described in terms of four shells at increasing distances from the centre: a 'small' tetrahedron, a 'large'

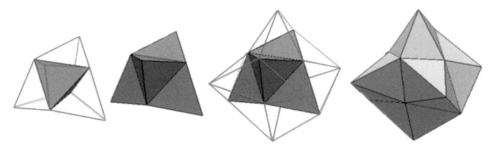

Figure 6.17 The building of a γ-brass cluster around a tetrahedral cluster. (a) Four tetrahedra placed on the faces of a central tetrahedron give (b) the *stella quadrangula*; (c) two more tetrahedra on each edge of the inner tetrahedron complete the rings of five tetrahedra round each edge and we have (d) a cluster of 17 nearly regular tetrahedra.

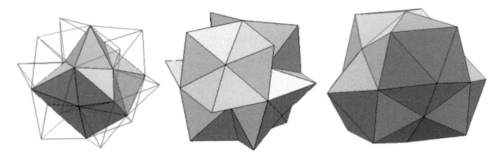

Figure 6.18 The completing the 'five-rings' around the edges of the secondary tetrahedra leads to a configuration of 41 tetrahedra and 26 vertices; sixteen more tetrahedra can be placed on this structure without changing the number of vertices. We arrive at the model of the 26-atom γ-brass cluster as *four interpenetrating icosahedral clusters*.

tetrahedron, an octahedron, and a cuboctahedron (with rectangular instead of square faces) (Bradley & Jones 1933).

The 52-atom cubic unit cell of the γ alloys contains two of these clusters. The same cluster can act as the basic building block in hexagonal structures (Sugiyama *et al.* 1998; Takakura *et al.* 1998; 1999).

In Figure 5.17, we see that the γ-brass cluster can be understood in terms of a packing of 41 tetrahedra. *Without increasing the number of vertices*, inserting 16 more tetrahedra reveals the structure to be *four interpenetrating icosahedra* sharing a common tetrahedral building block (Plate IX).

This cluster can be further augmented by placing three extra spheres over four of the triangular faces of the cuboctahedral shell. We then have 38 spheres. In the γ-alloys, these 38-atom clusters share atoms, joining up along the threefold axes to form a structure in which the cluster centres lie on a bcc lattice (Belin & Belin

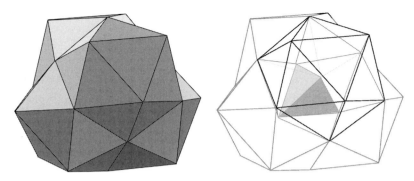

Plate IX The 26-atom γ-brass cluster as a cluster of four interpenetrating icosahedra.

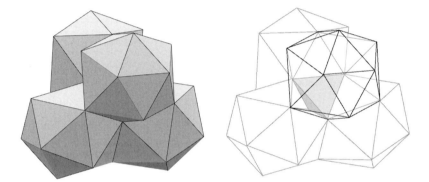

Plate X The Pearce cluster. Four icosahedra on the faces of a tetrahedron.

2000). The network of centres and bonds within the 38-atom cluster forms a structure rather like the Pearce cluster (1978) – four icosahedra in face contact with each other and with a central tetrahedron (Plate X) – except that the icosahedra are in this case centred, so they are are not precisely Pearce's 'oblate' icosahedra.

6.5 Clusters of Friauf Polyhedra

A Friauf–Laves phase is a Frank–Kasper phase with 12- and 16- coordinated atoms. It can be visualised in terms of the polyhedron packing Fd$\overline{3}$m: $3^3 + 3.6^2$ with 16-coordinated atoms at the centres and 12-coordinated atoms at the vertices of the truncated tetrahedra; pairs of neighbouring truncated tetrahedra sharing hexagonal faces are related to each other by inversion in the centres of the hexagons.

Centred truncated tetrahedra (Friauf polyhedra) can occur as building blocks in much more complicated structures. The 104-atom cluster (Figure 6.10) is an example. Figure 6.19 indicates the initial stages only of the building up of Samson's description of the highly complex aluminium–manganese β phase (Samson 1965).

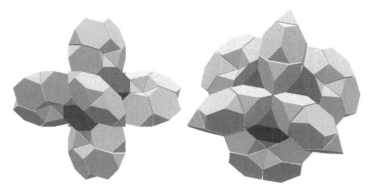

Figure 6.19 A cluster of Friauf polyhedra viewed along a twofold and along a threefold axis. Four Friaufs (dark grey) around a central tetrahedron, and six 'five-rings', each sharing an edge with the central Friauf.

Figure 6.20 illustrates the arrangement of the truncated tetrahedral 'building blocks' in α-Mn. The alpha phase of manganese has a large unit cell containing 58 atoms (Oberteuffer & Ibers 1970). It is a Frank-Kasper phase in which the arrangement of atoms is, for a pure metal, very intricate. The simplest description of its structure is in terms of the Friauf–Laves unit, consisting of a central atom coordinated to 14 others – 12 at the vertices of a truncated tetrahedron and four over the centres of its hexagonal faces. In α-Mn, pairs of neighbouring truncated tetrahedral units are related by *reflection* in a hexagonal or in a triangular face, rather than by inversion in the centres of the hexagonal faces. If these units are placed at the vertices of a bcc lattice, then each of them can be linked to the eight surrounding it by linkages in the form of similar units (four as in Figure 6.20a and four as in Figure 6.20b). Figure 6.20c indicates the unit cell structure and a further extension of the construction is shown in Figure 6.20d.

In each cubic unit cell of this structure there are thus two polyhedral units 'of type I' (those at the bcc positions) and eight of 'type II' (those that form the linkages). The central atoms of all units are 16-coordinated, though three of the four atoms located over the hexagonal faces of the type II units are *not* centres of other units, they are *vertices* of other units. Moreover, all the vertices of type I units are 12-coordinated – with coordination shells that are remarkably close to regular icosahedra. The type II units each have nine icosahedrally coordinated vertices and three 14-coordinated vertices (Sadoc & Mosseri 1999). The structure thus looks like that of a ternary alloy – though all the atoms are manganese! (The structure of β-Mn is also very complex – see Chapter 7. Why manganese atoms should interact with each other in such intricate ways is puzzling. . . .)

explained in the Lord and Ranganathan paper, the three small decagons in Figure 6.25 can in this case be interpreted as *towers of pentagonal antiprisms*, provided the large decagon in the figure is interpreted, not as the 'quasi unit cell' of Steinhardt *et al.*, but as a decagon smaller by a factor τ^{-1}. The towers are linked by tetrahedra rather than octahedra, as in Figure 6.28 (Steurer *et al.* 1993).

Figure 6.27 Li's (1995) model of decagonal Al–Mn constructed from a quasi unit cell. The structure consists of towers of icosahedra and pentagonal bipyramids, bonded by octahedral linkages.

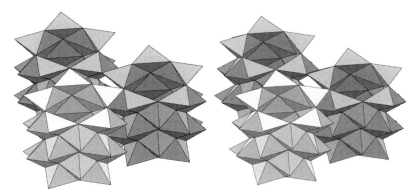

Figure 6.28 Atomic sites in the quasi unit cell model of Al–Ni–Co lie at the vertices of a structure of pentagonal antiprisms and tetrahedra (and – omitted in the figures – at the centres of the pentagons).

6.8 Triacontahedral Clusters

The ubiquitous presence of triacontahedra in the 3D analogues of the Penrose rhomb patterns was emphasised by Mackay at an early stage in the development of quasicrystal theory (Mackay 1987). In the same paper it was pointed out that models based on atomic clusters with icosahedral symmetry would provide a more realistic approach than the emphasis on tiling patterns.

Large triacontahedral clusters built around Bergman units are a striking feature of the 'R-phase'. An icosahedral quasicrystalline phase, the 'T2 phase' is known to be structurally closely related to the R-phase. Model building, in an attempt to understand the T2 phase, is therefore largely a matter of understanding how quasi-periodicity can arise from the combination of interpenetrating triacontahedra. In the model constructed by Audier & Guyot (1988) large triacontahedral clusters, identical to the R-phase cluster (Figure 6.12) are centred at the vertices of the Kowalewski units of the standard 3D tiling, of edge length τ^3 times the triaconta-hedron edge length.

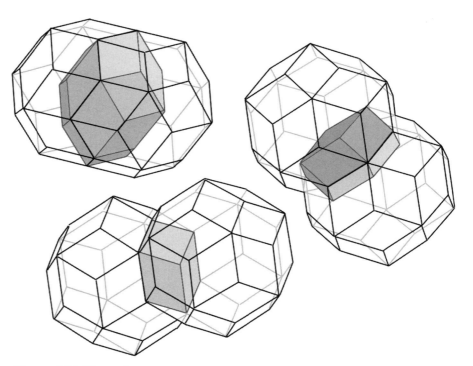

Figure 6.29 The various ways in which triacontahedra can interpenetrate, sharing vertices. The overlap region is a rhombic icosahedron, a rhombic dodecahedron, or an oblate rhombic hexahedron, according to whether the 'twinning' is along a five-, two- or a threefold axis. Along a twofold axis a pair of triacontahedra can also combine by sharing a single face.

In the model proposed by Lord *et al.* (2000; 2001a) the triacontahedral clusters are not centred at the vertices of the τ^3 Kowalewski units but at the *golden mean* positions on the edges of the units and on the long diagonal of the prolate units. The large triacontahedra then interpenetrate in the ways shown in Figure 6.29. The network of triacontahedron centres and the lines joining neighbouring triacontahedra along two-, three- and fivefold axes are given as follows (for triacontahedra of unit edge length):

Axis	Distance
2'	$2\tau^2$
2	2τ
3	$\tau^2\sqrt{3}$
5	$\gamma = \sqrt{(2 + \tau)}$

These lengths occur in the *Zometool* construction kit (Baer 1970; 1984; Booth 1992; Hart & Picciotto 2002), which employs icosahedral connectors to which rods can be affixed along two-, three- and fivefold axes. (The system is thus an icosahedral analogue of Pearce's 'Universal Node' construction system, based on cubic symmetry (Pearce 1978).) The rod lengths in the Zometool system are 1, $\sqrt{3}$, γ and multiples of these by powers of τ.

6.9 Crystalloids

An 'ideal' crystal is triply periodic; thus, by definition, it is infinite in extent. Real crystals have bounding surfaces. The conditions at or near the surface are not the same as those in the bulk material. The arrangement of atoms or molecules in small clusters will differ from the arrangement in the interior of a large crystal (Johnson 2002). A large number of investigations in recent years have addressed this problem. Alan Mackay coined the term 'crystalloid' (Mackay 1975): 'I would define a crystalloid as follows: A crystalloid is a configuration of a finite number of identical sub-units, assembled in a true free-energy minimum in a unique, regular and reproducible way into a cluster which is not recognisable as being a crystallite. A crystallite is a small crystal, distorted by the presence of a surface, but of recognisably the same structure as that which an infinite region of the same material would adopt. Crystalloids may contain exact or approximate symmetry operations which are not allowed in the 230 space groups.'

The question of the optimal packing of an infinite number of equal spheres is readily defined. Maximum density is an obvious criterion of optimality, and the optimal packing is the fcc arrangement (Kepler 1619; Hales 1996; 1997). For a finite number of spheres the concept of the density of the packing is ill-defined; other candidates for optimisation are potential energy of the cluster, for various choices of pair potential.

The first rigorous and systematic investigation of the problem was made by Hoare & Pal (1971), employing the Lennard-Jones potential – the simplest and most frequently used standard potential in computer simulations; specifically, $(\sigma/r)^6$ – $(\sigma/r)^{12}$. Minimal energy configurations can be described *approximately* in terms of the packing of equal spheres. Polytetrahedral structures predominate. For five atoms there are two stable minimal configurations: a trigonal bipyramid and a square pyramid (half of an octahedron). The latter has the lower energy. It is the only case of a Lennard-Jones cluster that is *not* polytetrahedral.

The number of different stable configurations increases with increasing number N of atoms.

The absolute minima for $N = 6, 7, 8$ and 9 turn out to be portions of Boerdijk–Coxeter helix! (built, respectively, from 3, 4, 5 or 6 tetrahedra). For $N = 13$ the absolute minimum is a centred icosahedron – the cuboctahedral arrangement is metastable.

Hoare and Pal examined growth sequences, starting with a 'seed' of given symmetry and adding atoms successively. For example, starting from the 7-atom pentagonal bipyramid the symmetry reappears at $N = 12, 13$ (icosahedron), 19 (double icosahedron), 24 and 33 (a dodecahedron around an icosahedron, as in the Bergman cluster). Starting from a tetrahedron ($N = 4$), tetrahedral symmetry is regained at $N = 8$ (the *stella octangula*, Figure 6.30), 26, 38 and 66.

Also considered by Hoare and Pal were close-packed structures – small pieces of cubic close packing (octahedra and tetrahedra) modified by the action of the potential. A curious phenomenon occurs for the 8-atom arrangement given by an octahedron with tetrahedra on two adjacent faces. The cluster turns into Bernal's dodecadeltahedron (Figure 4.9e).

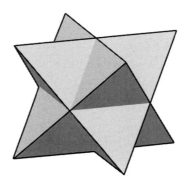

Figure 6.30 Kepler's *stella octangula*.

Many investigators have contributed to the large body of knowledge now available on 'small' clusters formed by various potentials. A very extensive data base on the Internet is the Cambridge Cluster Data Base. The Birmingham Cluster Data Base gives information on the work of the Birmingham group. The work of Jonathan Doye and his coworkers deserves particular mention; see for example Doye & Wales (2001) on polytetrahedral cluster (produced by a potential different from the Lennard-Jones potential), in which the occurrence of 'magic numbers' of atoms for particularly stable clusters is emphasised (an analogy with the 'magic numbers' of nucleons that produce particularly stable isotopes (Pauling 1965)), and for which the disclination networks are characterised by six-membered rings. Clusters of fullerene molecules (C_{60}) have also been studied (Doye & Wales 1996).

Manoharan *et al.* (2003) have obtained clusters of N equal spheres with characteristic shapes quite different from those predicted from the minimising of pair potentials, but with consistent shapes for each value of N. The clusters were obtained from uniformly sized spherical polymer particles adhering to the surface of a liquid drop; the spheres were caused to conglomerate by evaporating the drop. The sequence of cluster shapes so obtained was shown to correspond, up to $N = 11$, to minimisation of the second moment of the mass distribution – which, apart from an irrelevant numerical factor, is the sum of squares of the distances of the spheres from the centre of mass of the cluster. The configurations given by clusters of equal spheres with minimal second moment have been obtained theoretically by Sloane *et al.* (1995), up to $N = 35$.

The kind of clustering depends, as one might expect, on the particular condition that brings it about. There is quite a range of possible pair potentials that can be employed to simulate clustering. A survey of the results, with comparisons of the various types of clustering, has been given by Atiyah & Sutcliffe (2003).

7

Helical and Spiral Structures

A *helix* is a curve in three dimensions (3D) that winds around a circular cylinder, maintaining a constant angle with the direction of the cylinder's axis. An equiangular (or logarithmic) spiral is a curve in 2D that maintains a constant angle with the radius from a fixed point. The helix and the equiangular spiral are limiting cases of the *concho-spiral* or *conical helix*, which is a curve on the surface of a circular cone, obtained by projection on to the cone of an equiangular spiral in a plane perpendicular to the cone's axis. The property of *self-similarity* of these curves is responsible for their occurrence in all natural forms that grow by a process of steady accretion. In his classic work *On Growth and Form* D'Arcy Wentworth Thompson (Thompson 1917; 1996) considered the shapes of seashells, horns of sheep, goats and antelopes, and other biological examples, from this mathematical viewpoint. (The abridged 1992 edition omits the interesting chapter on *phyllotaxis* that discussed the occurrence of equiangular spirals and Fibonacci numbers in plant structure.) Less well-known is James Bell Pettigrew's *Design in Nature* (1908). A fascinating survey of spirals and helices in nature and art is Cook's *The Curves of Life* (1914; 1979). The so-called 'spiral' staircase is a familiar example of a human artefact of helical form – Cook illustrates many fine examples. Ghyka's book (1946; 1977) deals more briefly (and rather more fancifully) with the same topics.

7.1 Screw Axes

Helical structures are, of course, present in all crystalline materials whose space group symmetry contains screw transformations. The earliest attempts at classification of the internal structure of crystals did not recognise the important fact that a rotation in a point group symmetry (crystal class) could be a manifestation of a screw transformation, and that reflection symmetries in point groups could be the result of glide-reflections. Screw axes are of particular importance in the chiral space groups. We then have the property of enantiomorphism, where a crystalline

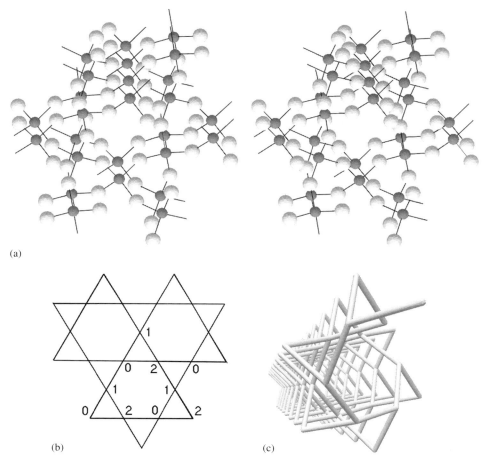

Figure 7.1 Three ways of representing the structure of β-quartz. (a) Stereo view of ball and stick model. Silicon atoms represented by black balls, oxygen by white balls, (b) view along the axis, oxygen atoms omitted. The numbers labelling vertices are z-coordinates in units of $c/3$, (c) network of rods representing Si—O—Si bonds. There are right-handed triangular helices and left-handed hexagonal double helices. The symmetry is P6$_2$22. The enantiomorphic form, of course, has symmetry P6$_4$22.

solid can exist in two forms, right-handed and left-handed. Figure 7.1 illustrates a well-known example, β-quartz.

7.2 Polyhedral Helices

The screw transformations associated with the screw axes of a triply periodic structure are of a limited kind, namely, those obtained by combining a translation along

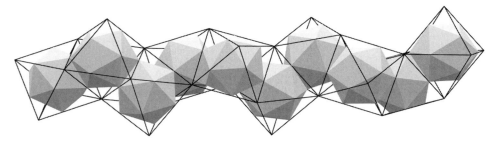

Figure 7.2 An icosahelix inscribed in an octahelix.

an axis with a rotation about the axis through an angle $2\pi/n$, where $n = 2, 3, 4$ or 6. Interesting geometrical structures can be generated by more general screw transformations. The Boerdijk–Coxeter (B-C) helix discussed in Section 4.5 is an example. The rotational advance per tetrahedron is $\cos^{-1}(-2/3)$, an irrational multiple of 2π. The structure is aperiodic. Lidin & Andersson (1996) defined a *regular polyhedral helix* to be a structure of face-sharing Platonic polyhedra generated from a generic polyhedral unit by a screw transformation. They enumerated the cases that do not lead to steric hindrance (and, consequently, self-intersecting structures). There is then a unique tetrahedral helix (the B-C helix), one cubic helix, two octahedral helices, two icosahedral and two dodecahedral. The octahedral helix whose octahedral vertices are the mid-points of the edges of a B-C helix, and the icosahedral helix obtained by inscribing icosahedra in the octahedra (as in Figure 7.2) were mentioned by Pearce (1978).

Any Euclidean transformation is uniquely determined if four non-coplanar points and their four images are given. See Lord & Ranganathan (2001b) and Lord (2002) for a simple matrix approach. Various extensions of the concept of a polyhedral helix are of interest, such as the twisted columns of interpenetrating icosahedra (Figure 7.3).

7.3 Polyhedral Rings

Among the structures of face-sharing regular polyhedra that can be generated by a single Euclidean transformation, some varieties that encounter steric hindrance are not without interest. In particular, in some of the cases where the generating transformation is a pure rotation or roto-reflection rather than a screw transformation the rotation angle turns out to be close to $2\pi/n$ for integer n. A slight deformation of the polyhedron will then produce a closed ring. The simplest cases are the ring of five face-sharing tetrahedra and the ring of three face-sharing dodecahedra. Further examples are given in Figure 7.4. The ring of 20 tetrahedra is generated

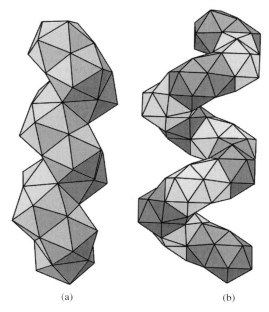

(a)   (b)

Figure 7.3 (a) A helix of interpenetrating icosahedra, (b) a helix generated by a fivefold screw transformation from a double icosahedron.

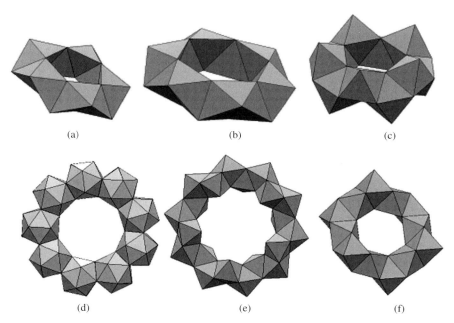

(a)   (b)   (c)

(d)   (e)   (f)

Figure 7.4 Examples of ring structures formed from almost regular polyhedra: rings of 20 tetrahedra, 24 tetrahedra and 10 octahedra, generated by rotary-reflection, a ring of 9 icosahedra generated by a rotation, a ring of 20 octahedra and a ring of 12 octahedra generated by the point group $\bar{3}2$. In this final example, the octahedra are regular.

from a trigonal bipyramid by a tenfold rotary-reflection. The ring surrounds a void in the form of a pentagonal antiprism. The dihedral angles θ and θ_T of a pentagonal antiprism and of a tetrahedron satisfy $\theta + 3\theta_T = 349.8°$ so the fit is nearly perfect. For the ring of 24 tetrahedra it is even closer; $\theta + 3\theta_T = 359.2°$. For a ring of nine regular icosahedra there is a deficit angle of only $1.8°$ between faces. A ring of 96 'nearly regular' tetrahedra, generated by a sixteenfold rotation applied to a six tetrahedron portion of B-C helix, was discovered by Walter (2000); it is discussed in Lord (2002).

7.4 Periodic Tetrahelices

The B-C helix is aperiodic – that is, it has *no translational symmetry* because the screw transformation that generates it has rotation angle θ, where $\cos(\theta) = -2/3$, which is an irrational multiple of 2π. A slight twisting of the structure can produce a periodically repeating tower of very slightly distorted tetrahedra. The rational approximants to $\cos^{-1}(-2/3)/2\pi$ are 1/3, 3/8, 4/11, 11/30, etc. For example, $8\theta \sim 1062.4°$ is close to $3.360° = 1080°$, so that a periodic structure with 8 tetrahedra in the repeat is obtained by a twist amounting to only about $2.2°$ per tetrahedron. Figure 7.5 shows the view along the axis. A further deformation, obtained by moving the odd numbered vertices in this figure slightly closer to the axis, produces a tetrahelix with the property that three of them can be placed in face contact, with the three axes orthogonal (Figure 7.6). Thus, a *rod packing* (of the type shown in Figure 4.15c) can be produced in which the rods are tetrahelices instead of cylinders.

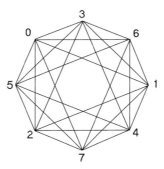

Figure 7.5 A periodic tetrahelix. Numbers denote the *z*-coordinate.

Analysis of the structure of β-Mn has revealed that the manganese atoms lie at the vertices of this remarkable structure (Nyman *et al.* 1991). Considered as a packing of spheres, two sphere sizes are involved, so β-Mn may be regarded as a binary alloy – though both kinds of atom are manganese! Figure 7.7 gives a stereo pair of images. Observe, incidentally, that further tetrahedra can be inserted into angles in the structure without adding more vertices.

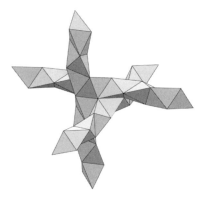

Figure 7.6 Three orthogonal tetrahelices in face contact.

Figure 7.7 The structure of β-Mn.

7.5 A Periodic Icosahelix

The structure shown in Figure 7.3b is generated by a fivefold screw transformation. A slight twist will produce a periodic structure generated by a sixfold screw transformation – six of the double-icosahedron units in the repeat. The vertices of this structure, including the points at the centres of the component icosahedra, define two interpenetrating structures; a periodic helix of (non-interpenetrating) vertex-sharing icosahedra and a periodic helix of (non-interpenetrating) face-sharing icosahedra. The structure is thus very close to the description of a highly complex cobalt–zinc phase analysed and described by Boström & Lidin (2002). The basic structure is a double helix of vertex-sharing Zn icosahedra surrounding Co atoms as in Figure 7.8. The helices repeat after six icosahedral units. These double helix structures pack together by face-sharing, as in Figure 7.9, to produce the chiral crystalline phase, with symmetry $P6_2$. These icosahedra are virtually perfectly regular. The face-sharing icosahedra that constitute the helices that interpenetrate those

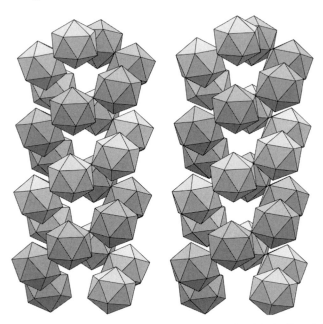

Figure 7.8 A stereo pair of images illustrating the double helix of icosahedra occurring in δ-Co_2Zn_{15} (Boström & Lidin 2002).

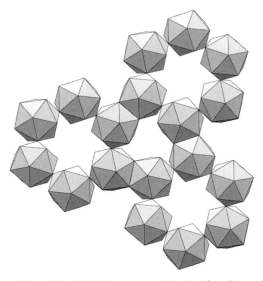

Figure 7.9 View along the sixfold screw axis showing how the helices pack together by face-sharing.

Figure 7.14 (a) 8 anticlockwise spirals and 13 clockwise spirals in a pinecone. Photo copyright Istvan Hargittai, reproduced with his kind permission, (b) the florets of a daisy, arranged in 21 anticlockwise and 34 clockwise spirals. Photo by courtesy of Radoslav Jovanovic, (c) a sunflower head with a (34, 55, 89) pattern from Thompson's *On Growth and Form* (1917; 1942). Reprinted with permission of Cambridge University Press.

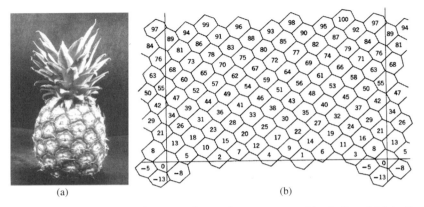

(a)   (b)

Figure 7.15 (a) The subunits on a pineapple are arranged in 5, 8 and 13 helical rows. Photo reproduced by kind permission of Radoslav Jovanovic, (b) the Dirichlet regions of the points of a 'Fibonacci helix'. From Coxeter's *Introduction to Geometry*. Reprinted with permission of John Wiley and Sons Inc.

(5, 8, 13) pattern will change to an (8, 13, 21) pattern, and so on. (This can be appreciated by noting that the hexagons adjacent to hexagon 0 in Figure 7.15 are hexagons 5, 8 and 13, and that point 21 will move closer to point 0 as d is decreased until the hexagons turn into rectangles; thereafter, hexagon 21 will be in contact with hexagon 0 and hexagon 5 will not.)

Coxeter showed that we get the (F_{n-1}, F_n, F_{n+1}) pattern if:

$$d = 2\pi/F_n\tau^n.$$

The question then is, why should *this particular angle*, $2\pi\tau$ (equivalently $2\pi/\tau^2$), occur in plant structure? A clue comes from looking at the form of the corresponding *Fibonacci spiral* or *golden spiral* arrangement of points in a plane given by the sequence of polar coordinates:

$$(r_n, \theta_n) = (\sqrt{n}, 2\pi n/\tau^2)$$

(Figure 7.16). The prescription $r_n = \sqrt{n}$ ensures that the first n points lie within a circle of radius n, for any n, so that the Dirichlet regions will be of nearly equal area (Figure 7.17). (In real plant forms exponential growth also contributes to the pattern, so that the spirals are closer to logarithmic spirals, as in the daisy in Figure 7.14b.)

The striking feature of the arrangement of points in the Fibonacci spiral arrangement is the *uniformity* of their distribution. It clearly corresponds to a very *efficient packing of identical subunits*. Michael Naylor has given a very clear explanation of

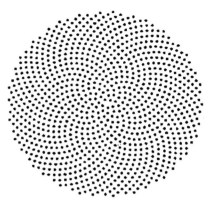

Figure 7.16 The Fibonacci spiral to $n = 1000$.

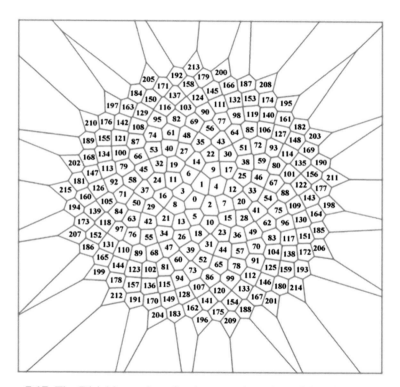

Figure 7.17 The Dirichlet regions for the central portion of the Fibonacci spiral.

this property of the golden spiral (Naylor 2002). The brief treatment we present here should be supplemented by consulting this work. See also Marzec & Kapraff (1983) and Kaproff (1992). The uniformity of the spiral distribution of points produced by the golden angle $2\pi/\tau^2 \sim 137.508°$ is more striking when compared with

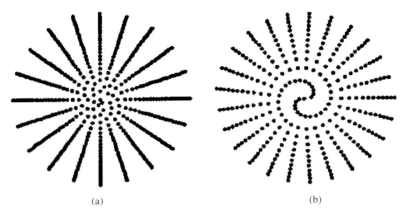

(a)   (b)

Figure 7.18 The distribution of points $(r, \theta) = (\sqrt{n}, n\alpha)$ for: (a) $\alpha = 3/20$, giving 20 radial arms – the 7 spirals apparent in the center correspond to the fact that 3/20 is close to $3/21 = 1/7$; (b) $\alpha = 12/25$, giving 25 radial arms, and two spirals in the center corresponding to the closeness of 12/25 to $12/24 = 1/2$. Figures from Naylor (2002), reprinted with permission of Michael Naylor.

Figure 7.19 The π spiral to $n = 10,000$. Figure from Naylor (2002), reprinted with the author's permission.

the patterns produced by other angles $2\pi\alpha$. If α is rational, $\alpha = M/N$, then the $(n + N)$th point will lie on the same radius as the nth point and we get a pattern of N radial lines, as in Figure 7.18.

Figure 7.19 shows the distribution obtained by taking $\alpha = \pi$. Note the seven spirals in the center of the pattern, corresponding to the well-known rational approximation 22/7. The succession of increasingly accurate approximants to an irrational number is given by the cut-offs of the continued fraction expression for the number:

$$a_0 + \cfrac{1}{a_1 + \cfrac{1}{a_2 + \cfrac{1}{a_3 + \cdots}}}$$

For π, $\{a_0, a_1, a_2, a_3, \ldots\} = \{3, 7, 15, 1, 292, 1, \ldots\}$. The next approximants after 22/7 are 333/106 and 355/113. The 113 spirals look almost straight in the figure. The 106 spirals are not apparent – they are dominated by the 113 before they have a chance to develop.

The reason for the very uniform distribution given by the golden angle is that, as mentioned in Section 2.8, in the continued fraction expression for τ the integers $a_0, a_1, a_2, a_3, \ldots$ are all 1, so that the sequence of approximants $(1, 2/1, 3/2, 5/3, 8/5, \ldots)$ converges more slowly than for any other irrational number (τ is, in this sense, the 'most irrational' of all irrational numbers!)

A much deeper understanding of how these patterns – and patterns in nature generally – arise is provided by dynamical theories of morphogenesis, in which the

Figure 7.20 Spiral phyllotaxis in a cactus, possibly induced by the buckling pattern of its curved surface. Photo copyright Magdolina Hargittai, reprinted with her kind permission.

interaction between chemical morphogenesis and their relative diffusion rates is expressed in terms of differential equations. Alan Turing (1952) was a pioneer in the exploration of this approach. He was particularly fascinated by spiral Fibonacci type phyllotaxis. A survey of Turing's later – unpublished – work on phyllotaxis has been provided by Jonathan Swinton (2004).

Shipman & Newell (2004) have obtained the same spiral arrangements by simulating the buckling of the convex surface of an elastic material that grows at a slower rate than the outer shell. They obtained a variety of 3D patterns strongly resembling the forms of various species of cacti with spiral morphologies – for example, Figure 7.20. Nanoparticles with a very striking structure have been produced by Li *et al.* (2005). They were produced by cooling a mixture of silver oxide and silicon oxide below the melting temperature of silver. Under carefully controlled cooling conditions spheres of silver with diameters of only a few microns were formed, with remarkably uniform distributions of SiO spherules on their surfaces. Conical forms exhibiting spiral Fibonacci patterns, closely resembling the spiral structures of plant forms, were also obtained.

7.10 Spiral Distribution of Points on a Sphere

In Section 4.8 we drew attention to the problem of distributing N points uniformly over the surface of a sphere. Except in a few very special symmetrical cases, such as the icosahedral arrangement for $N = 12$, the particular optimal solution depends on the specific criterion of 'uniformity'. An extensive literature dealing with this problem area exists. An interesting article on various approaches to the 'uniform distribution' problem when N is large has been given by Saff & Kuijlaars (1997). Though, in general, different criteria give rise to different optimal configurations for a given N, for large N the points tend to distribute themselves in hexagonal arrays, necessarily disrupted to fit the curvature of the spherical surface. Rakhmanov *et al.* (1997) described a simple algorithm that distributes N points 'uniformly' in a spiralling pattern with the property that the configuration resembles a hexagonal lattice over much of the spherical surface. They demonstrated that the point distributions obtained from their algorithm give good minimal energy solutions for N points on a sphere under the influence of a pair potential $1/\log(r)$. That these spiralling patterns of points on a spherical surface provide effective answers to the Tammes problem was first shown by Székely (1974).

We shall present here an interesting variant on the algorithm that is closely related to the algorithm that we met in the previous section, that generates a Fibonacci spiral pattern in the plane (Figure 7.16).

Observe that a strip cut from a plane tiled by equilateral triangles can be rolled around a cylinder so that the three families of triangle edges become three helices,

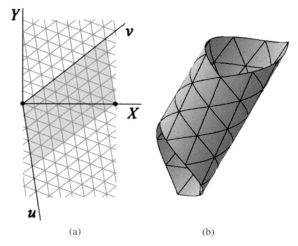

Figure 7.21 (a) A portion of the plane hexagonal lattice in the CHL (3, 5, 8) the point [0, 0] and the point [5, 9] are identified; (b) a portion of a CHL of type (3, 5, 8). The construction is essentially the same as that shown in Figure 7.10 except that the plane lattice is in this case rolled to form a cylinder, instead of folded.

winding around the cylinder and intersecting each other at 60°. We call the resulting set of vertices a *cylindrical hexagonal lattice* (CHL) of type (l, m, n) if the three families contain l, m and n helices, respectively. Figure 7.21 illustrates the case (3, 5, 8). Without loss of generality we may assume $0 \leqslant l \leqslant m \leqslant n = l + m$ (Lord 2002).

The points of the plane hexagonal lattice can be labelled by a pair of *integer* indices $[u, v]$, referred to a pair of oblique axes subtending 120°. A little trigonometry establishes that the Cartesian coordinates (X, Y) of the lattice point $[u, v]$ are given by:

$$X = u \cos \alpha + v \sin(\alpha - \pi/6), \quad Y = u \sin \alpha - v \cos(\alpha - \pi/6)$$

where α is the angle between the X-axis and the u-axis:

$$\cos \alpha = (2m - n)/2p, \quad \sin \alpha = \sqrt{3}n/2p, \quad \text{where } p = \sqrt{(m^2 + n^2 - mn)}$$

(see Figure 7.21a). Thus,

$$X = \{(2u - v)(2m - n) + 3vn\}/4p, \quad Y = \sqrt{3}(un - vm)/2p.$$

When the lattice is wrapped around a circular cylinder with circumference p we get a CHL of type (l, m, n). A point $[u, v]$ of the lattice becomes a point on the cylinder with cylindrical coordinates (ρ, φ, z). Choosing the scale so that the cylinder has unit radius ($\rho = 1$):

$$\varphi = 2\pi\{(2u - v)(2m - n) + 3vn\}/4p^2, \quad z = \pi\sqrt{3}(un - vm)/p^2.$$

If m and n have no common factor, there are two points $[\mu, \nu]$ on the CHL with smallest absolute value of the z-coordinate (other than $z = 0$). They are given by the solutions of:

$$\mu n - \nu m = \pm 1$$

(which can readily be found from $\nu l \pm 1 = 0$ modulo m). It is immaterial which sign is chosen – all the points of the CHL lie on *a single generating helix* through $[0, 0]$ and $[\mu, \nu]$. (If m and n have a common factor d, then there will be d generating helices (Székely 1974).) A simple algorithm for generating a CHL is then:

$$\varphi_k = 2\pi kA, \; z_k = kh, \; (k = \ldots -2, -1, 0, 1, 2, 3, \ldots),$$

where

$$A = \{(2\mu - \nu)(2m - n) + 3\nu n\}/4p^2, \; h = \pi\sqrt{3}/p^2.$$

A distribution of points on the surface of a unit sphere centered on the axis of a cylinder of unit radius can now be obtained by the mapping that takes each point k on the sphere to be the intersection of the sphere by the perpendicular to the axis from the point k of the CHL on the cylinder. If the number of points on the sphere is sufficiently large the distribution so obtained will closely approximate to a plane hexagonal lattice in the 'equatorial' regions. This mapping from the cylinder to the sphere is an *equal area mapping* – if a sphere is cut into parallel slices of equal thickness, then the area of the portion of spherical surface is the same on all of the slices. Thus the perfect uniformity of the point distribution on the cylinder ensures 'uniformity' in the distribution over the surface of the sphere in the sense that the points on the sphere will occupy territories of equal area.

The sphere can have a point at each of the 'poles' if the value of h is changed slightly so that $2/h$ is an integer. Then $N = 1 + 2/h$ is the total number of points on the sphere. In other words, instead of taking h to be $\pi\sqrt{3}/p^2$, we may choose N to be an integer close to $1 + 2p^2/\pi\sqrt{3}$ and take $h = 2/(N + 1)$.

Interesting cases arise when m and n are *two successive terms in the Fibonacci sequence*. The pattern on the sphere in the regions around the poles then closely approximates to the *golden spiral* (Figure 7.16). A few of the 'best' values for N are 48 ($m = 6$), 125 ($m = 13$), 326 ($m = 21$), 850 ($m = 34$), 2225 ($m = 55$), … The distribution of 850 points on a sphere obtained in this way is illustrated in Figure 7.22.

Observe that, because of the identity

$$F_{r-1}F_r - F_{r-2}F_{r+1} = (-1)^r,$$

when $(l, m, n) = (F_{r-1}, F_r, F_{r+1})$, we can take $(\mu, \nu) = (F_{r-2}, F_{r-1})$. From the fact that F_{r+1}/F_r closely approximates the golden number τ for sufficiently large r, it

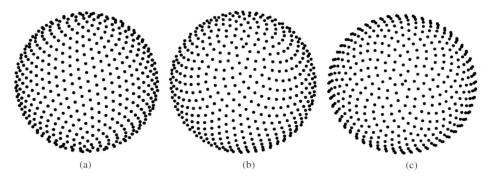

(a)   (b)   (c)

Figure 7.22 Spiral distribution of 850 points on a sphere: (a) view perpendicular to the axis. Note the hexagonal lattice arrangement in the region close to the 'equator'; (b) view at 60° to the axis; (c) view along the axis showing the Fibonacci spiral configuration in the polar region.

then follows that A is approximated by τ^{-2}. Therefore, for large N, A can be replaced by τ^{-2} without making a perceptible difference to the point distribution (e.g. for $m = 34$, $A = 0.381669\ldots$ whereas $\tau^{-2} = 0.381966\ldots$). This accounts for the golden spiral configurations around the poles.

8

3D Nets

8.1 Infinite Polyhedra

Associated with every vertex of a 2-dimensional (2D) tiling, a polyhedron, or a 3D net (or their higher-dimensional analogues) is a *vertex figure*. Its vertices are all the vertices connected by an edge to the given vertex. A regular polyhedron in Euclidean 3 space (E_3) is a polyhedron all of whose faces are regular p-gons and all of whose vertex figures are regular q-gons. There are just five regular polyhedra: the five 'Platonic solids', denoted by $\{p, q\}$.

Coxeter and Petrie (Coxeter 1937) discovered three more, simply by relaxing the definition of a *regular* polygon. If the vertex figures in the above definition of a regular polyhedron are not planar, but regular *skew* polygons whose vertices are related to each other by roto-reflection rather than rotation, then there are three more regular polyhedra $\{4, 6\}$, $\{6, 4\}$ and $\{6, 6\}$ (Figure 8.1). Unlike the five well-known Platonic solids, these three polyhedra are infinite. They are triply periodic; their symmetry

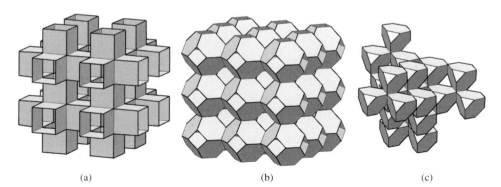

(a)   (b)   (c)

Figure 8.1 Portions of the three infinite regular polyhedra: (a) $\{4, 6\}$. Each labyrinth consists of 50% of the cubes of the space-filling $Pm\overline{3}m$: 4^3. (b) $\{6, 4\}$, the dual of $\{4, 6\}$, each labyrinth consists of 50% of the space-filling by truncated octahedra. The 4-connected net consisting of its vertices and edges is the zeolite framework SOD (sodalite). (c) $\{6, 6\}$. Labyrinths consist of 50% of the tetrahedra and truncated tetrahedra of the space-filling $Fd\overline{3}m$: $3^3 + 3.6^2$.

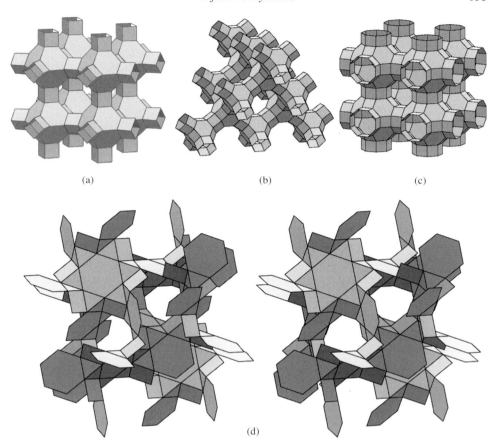

(a)   (b)   (c)

(d)

Figure 8.2 Examples of infinite Archimedean polyhedra. (a) The zeolite framework LTA is the net of vertices and edges of the space-filling Im $\overline{3}$m: $4^3 +$ $4.6^2 + 4.6.8$. One labyrinth of the polyhedron $4^2.6^2$ consists of the cubes and truncated octahedra and the other consists of the truncated cuboctahedra. (b) A polyhedron $4^3.6$ constructed from cuboctahedra and hexagonal prisms gives the faujasite network (zeolite framework FAU). Note the large 'pores' or 'cages' $\{4^{18}.8^4.12^4\}$ accessed through their 12-rings. (c) A different $4^3.6$, with congruent labyrinths, obtained from the space-filling Im$\overline{3}$m: $4^2.8 + 4.6.8$. (d) A stereo pair of images of a polyhedron $3.4.3^2.6$ with the same symmetry (Ia $\overline{3}$d) and topology as the 'gyroid' minimal surface.

groups are space groups instead of point groups. They have the interesting property that their surfaces partition E_3 into two congruent labyrinthine regions.

Contemplation of these three objects provides an interesting introduction to the intricate problem of investigating and classifying 3D nets. (They also serve as an introduction to the surfaces to be dealt with in the next chapter.) Their vertices and edges constitute examples of uninodal nets. Extending the Coxeter–Petrie idea further, it is natural to consider the analogous generalisations of the Archimedean

solids and their duals (Wachman *et al.* 1974). A few examples of such infinite polyhedra are illustrated in Figures 8.2 and 8.3.

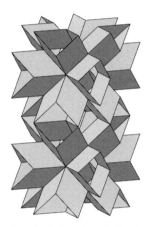

Figure 8.3 Two unit cells of an infinite rhombohedron with the same symmetry
and topology as the minimal surface C(P) (see Chapter 9).

8.2 Uniform Nets

The classic work on triply periodic nets in E_3 is Wells' *Three Dimensional Nets
and Polyhedra* (1977). This was the culmination of 20 years of investigations,
summarising the results published by Wells in *Acta Crystallographica* between
1954 and 1976, under the title 'The geometrical basis of crystal chemistry, parts
1–20'. It also contained some previously unpublished results. Some of the material
had already been included in Wells' earlier (1962) book. New 3-connected nets
were described by Wells in a much later publication (Wells 1983).

Wells introduced the concept of a *uniform* 3D net. In a uniform net of type (p, q)
all vertices are q-connected and the shortest circuits are p-gons. If the shortest circuit at *every* vertex is a p-gon, we have a topological generalisation of the concept
of a regular polyhedron. Wells found one uninodal (12, 3) net, seven (10, 3)s, three
(9, 3)s, fifteen (8, 3)s and four (7, 3)s. Wells' (10, 3)-a was mentioned in Chapter
3. It has cubic symmetry $I4_132$ or $I4_332$. The net (10, 3)-b is tetragonal (Figure 8.4).

If *regularity* of a tiling in E_3 is defined by analogy with regularity for polyhedra
and polytopes, that is, regular cells and vertex figures, then only the filling of E_3
by cubes is regular. However, if interest is focussed on the *net* formed by the vertices
and edges of the tiling, a less stringent definition of regularity is appropriate. A regular net in E_3 can be defined as a uninodal net whose vertex figure is a regular polygon or polyhedron (Delgado-Friedrichs *et al.* 2003). There are then just five varieties.

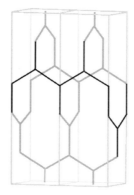

Figure 8.4 Two unit cells of Wells' uniform net (10, 3)-b.

8.3 Rings and Coordination Sequences

A triply periodic 3D net can be characterised by various *topological invariants*. Every such net has a *topological symmetry group* (TSG) consisting of the set of all permutations of its vertices that also map edges to edges. (The space group acts on the actual configuration in E_3 and is required also to preserve lengths and angles; in general it is a subgroup of the TSG.) The number of sets of vertices that are inequivalent under the TSG is an obvious example of a topological invariant – thus we can refer to the net as being uninodal, binodal, trinodal, etc. The number of inequivalent edges, of course, is also a topological invariant. Other obvious topological invariants are the connectivities of the vertices.

Wells introduced the invariants x and y. For any given vertex, x is the number of *rings* (circuits of minimal length) that contain the vertex, and for any edge, y is the number of rings that contain that edge. Figure 8.5 shows a portion of the (uninodal, 4-connected) diamond net (D net for short) with 29 vertices, consisting of all the

Figure 8.5 A portion of the D net, displayed as a packing of saddle polyhedra.

rings that contain a given vertex. (The subunit of a net consisting of all the rings that contain a given vertex is a *local cluster*. These clusters can be quite extensive (Hobbs *et al.* 1998).) Note that 12 hexagonal rings contain the central vertex, so that $x = 12$. Careful inspection of the figure also reveals that $y = 6$. For Wells' (10, 3)-a (Figure 4.5) the rings are all decagons, $x = 15$ and $y = 10$ (the highest values for any 3-connected net). 'Careful inspection of the figure' is not always an easy or reliable method; ring-finding algorithms have been devised (Goetzke & Klein 1991).

For each symmetrically distinguishable vertex of a net, one can associate a *vertex symbol* summarising information about the rings that contain the vertex. For a 4-connected net, let the symbols 1, 2, 3 and 4 label the four edges at a chosen vertex. For each of the six pairs of edges one can write down the size of the smallest ring containing the pair, and write as a subscript the number of such minimal rings. We obtain a symbol such as $3.4.8_3.9_4.8_3.9_4$, indicating that the edge pairs 14, 23, 24, 31, 34, 12 have, respectively, one 3-ring, one 4-ring, three 8-rings, four 9-rings, etc. These symbols are used extensively in the literature of 4-connected nets and are given for every vertex type in each of the known zeolite frameworks, in the *Atlas of Zeolite Framework Types* (Baerlocher *et al.* 2001). Clearly, the order of the three substrings that constitute the string AaBb.CcDd.EeFf is irrelevant, as also is the order of the two symbols (such as Aa and Bb) in each substring. A *unique* vertex symbol associated with a vertex of a 4-connected net can be obtained by taking the permutation that gives the smallest integer, when AaBbCcDdEeFf is interpreted as a number (in a sufficiently large base) (O'Keeffe & Brese 1992; O'Keeffe & Hyde 1997).

The vertex symbol for the D (diamond) net, for example, is $6_2.6_2.6_2.6_2.6_2.6_2$; those for the (binodal) ice XII net are $7_2.7_2.7_2.7_2.8_4.8_4$ and $7.7_3.7_2.7_3.8_4.8_4$ (O'Keeffe 1998a, b).

For any net vertex, one can define a *coordination sequence* (Meier & Moeck 1979). The first number n_1 of the sequence is the connectivity of the vertex; n_j is the

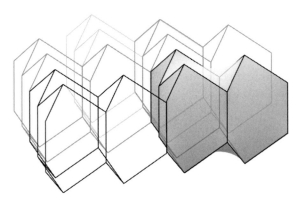

Figure 8.6 A portion of the 'lonsdalite' or 'wurzite' net. Lonsdaleite is a peculiar from of diamond, found in meteorites; wurzite is a form of zinc blende, ZnS.

number of distinct vertices that are connected to it by paths whose minimal 'length' (i.e. number of edges) is j. The coordination sequence for the D net, for example, is 4, 12, 24, 42, 64, 92, …; that of the closely related net shown in Figure 8.6 is 4, 12, 25, 44, 67, 96, …. Algorithms exist for computing coordination sequences from the data specifying a net. The sequences are valuable, in the context of classification, for distinguishing differences between nets that might otherwise be confused.

8.4 Vertex-Connected Tetrahedra

The basic structural unit for a silicate or an aluminosilicate (Liebau *et al.* 1986; Higgins 1994) is a silicon (or aluminium) atom coordinated to four oxygen atoms at the vertices of a regular tetrahedron. Each such tetrahedron can thus be joined to four others, by vertex-sharing. Since the bond angle of the bivalent oxygen atoms can vary quite widely (around an average of about 140°), a very large variety of configurations can arise. A very simple example is illustrated in Figure 8.7, another is β-quartz (Figure 7.1).

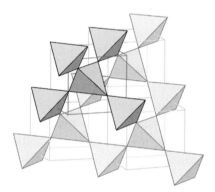

Figure 8.7 A unit cell of β-crystobalite, a silicate with symmetry group Fd $\bar{3}$m. A D net of SiO_4 tetrahedra.

The zeolites (Dyer 1988; Baerlocher *et al.* 2001) are essentially aluminosilicates in which the net of tetrahedra is a structure with large tunnels and voids (cages) in the tetrahedral framework, that can be occupied by other atoms or small molecules; hence their importance for the chemical industry. One of the simplest zeolites is sodalite; the structure of which is shown in Figure 8.8. Zeolites belong to the class of materials known as *clathrates*. Another important class of materials with open cage-like structures are the gas hydrates. Chlorine hydrate, for example, has a 4-connected network of water molecules corresponding to the vertices and edges of Figure 3.23. The chlorine atoms occupy the large 14-faced cages. These apparently esoteric crystal structures, elucidated in the 1950s by Linus Pauling are, in fact, of immense importance for the fate of human civilisation. At the bottom of the sea

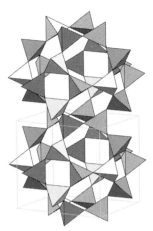

Figure 8.8 Two unit cells of sodalite. The dark grey and light grey tetrahedra are SiO_4 and AlO_4 units, which alternate. The large 'cages' contain Cl and Na atoms (not shown).

methane molecules are locked up in an ice-like structure of water molecules. This amounts to a huge reservoir of hydrocarbons, comparable to the world oil resources. There are, of course, difficulties in recovering the methane gas for burning. In Russia there are large resources of methane clathrates in the permafrost areas which can be excavated. However, the methane may be released by changes in temperature and pressure, and would affect the world climate.

A triply periodic arrangement of vertex-connected tetrahedra, with p distinct sets of tetrahedra distinguishable under the action of the space group symmetry, is completely specified if, for each of the p-types of tetrahedra, we specify which type is attached to each of the four vertices, and its orientation relative to the given tetrahedron. This approach to the storing of the data for these structures, and the reconstruction of the structure from the data, was proposed and investigated by Hobbs *et al.* (1998). Orientations could be specified, for example, by orthogonal matrices, by Euler angles, or by successive rotations about three coordinate axes. Hobbs *et al.* chose the latter method.

As an example consider the structure of β-quartz (Figure 7.1), which is uninodal ($p = 1$). Take a generic tetrahedron in a standard orientation and label its vertices (e.g. 0 (–1 –1 1), 1 (–1 1 –1), 2 (1 1 1), 3 (1 –1 –1)). The entire structure can be reconstructed from the information contained in Figure 8.9a. Thus, the vertices 0, 1, 2, 3 coincide, respectively, with the vertices 1, 0, 3, 2 of the neighbouring tetrahedra, and these four neighbouring tetrahedra are rotated from the standard orientation by rotations 60°, –60°, 60°, –60°, respectively, about the z-axis. Thus the structure of β-quartz is encoded in the symbol string 1 0 3 2 60 –60 60 –60. Similarly, crystobalite (Figure 8.7) is 1 0 3 2 90 –90 90 –90 (Figure 8.9b). For a p-nodal tetrahedral structure, p such symbol strings would be required, and in each,

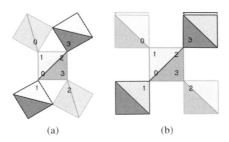

Figure 8.9 Basic units for uninodal tetrahedral structures: (a) β-quartz and (b) β-crystobalite.

the four types attached to the four vertices would be specified. We refer the reader to the original report of Hobbs *et al.* for further details and examples.

The method is an interesting example of a model-building algorithm based on self-assembly according to purely *local* rules. The method is, clearly, capable of generalisation to deal with more general network structures.

8.5 4-Connected Nets

The 4-connected nets have a very special place in crystal chemistry. The tetrahedral structures mentioned in the foregoing sections can be very simply represented by neglecting the vertices of the tetrahedra (the oxygen atoms) and joining their centres by straight edges. For example, the sodalite structure (Figure 8.8) is associated with the 4-connected net formed by the edges and vertices of the space-filling by truncated octahedra. This mode of representation is employed in the *Atlas of Zeolite Framework Types* (Baerlocher *et al.* 2001). The 'atlas' is a database prepared and continuously updated by the Structural Commission of the International Zeolite Association (IZA-SC) and is available on the Internet. The project began in 1970 with details on the 27 zeolite types known at that time; it now contains over 130 detailed descriptions of the nets associated with naturally occurring and artificially synthesised zeolites. Each is identified by a three-letter code – an abbreviation of the longer name of the zeolite.

The atlas gives descriptions of how the nets can be built up from subunits (secondary building units or SBUs). To describe 'zeolite-A' (LTA), for example, we can take a truncated octahedron as SBU, and link these units together by edges along [100] directions, to arrive at the 4-connected net shown in Figure 8.2a. Similarly, the faujasite (FAU) framework is obtained by linking the same SBUs along [111] directions – Figure 8.2b. The framework for MTT is the net constituted by the vertices and edges of the packing of dodecahedra and 16-hedra shown in Figure 3.22.

The zeolite frameworks mentioned above are all uninodal nets. Fourteen of the nets of known zeolites are uninodal. The number of symmetry-equivalent nodes in zeolite frameworks range from 1 to 16 (the highest values being 12 for the synthetic zeolite ZSM-5 (framework MFI) and 16 for STT). The variety and intricacy of zeolite structures is astonishing.

The atlas, of course, is concerned only with those 4-connected nets that occur as descriptions of known zeolites. From a more abstract theoretical viewpoint it is of course of interest to investigate and, if possible, to classify all possible triply periodic 4-connected nets and, for potential applications in the synthesis of new materials, to identify those that are 'plausible' as nets associated with clathrate structures. This, of course, was the motivation behind Wells' pioneering work. O'Keeffe & Brese (1992) gave detailed information (symmetries, nodal positions, coordination sequences and densities) on twenty-four 4-connected uninodal nets without 3- or 4-rings, of which 16 were previously unknown, and O'Keeffe (1992) gave similar descriptions of nineteen 4-connected uninodal nets with 3-rings, 'many of which we believe to be new'.

The crystalline structures of the various phases of *ice* are also describable in terms of 4-connected nets, since each water molecule is connected by hydrogen bonds to four others. The vertices of the net are occupied by the oxygen atoms; the hydrogen nuclei (protons) lie on the net edges (Runnels 1966). Most of the phases are hexagonal, the commonest form of ice corresponds to the net shown in Figure 8.6.

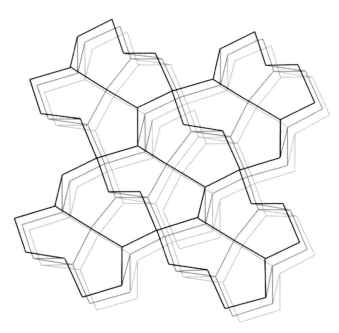

Figure 8.10 The 4-connected net describing the structure of ice XII, viewed along its tetragonal axis. It is composed of 7-rings and 8-rings.

Ice XII (Figure 8.10) is an unusual metastable tetragonal phase in which the smallest rings are heptagons (O'Keeffe 1998a).

The clathrate hydrates, like ice, are also based on 4-connected nets of H_2O molecules hydrogen-bonded to each other; in these materials there are large 'cages' that are occupied by atoms or small molecules (Pauling 1960). The net corresponding to chlorine hydrate, for example, is the 4-connected net of vertices of the polyhedral packing we showed in Figure 3.23. The chlorine molecules occupy the 14-hedra – they are too large to fit into the dodecahedra, which tend to be occupied by extra water molecules.

8.6 Crankshafts, Zigzags and Saw Chains

A large number of important 4-connected 3D nets consist of *layers* of 3-connected or (3, 4)-connected 2D nets, which may be either flat or puckered. The 3D structure arises from the edges interconnecting each layer with those above and below it. As a very simple example, we consider the 2D honeycomb tiling pattern of Figure 8.11. A vertex marked by a black circle is connected by an edge to the corresponding vertex in the layer above; one marked by a white circle is connected to the layer below. The resulting 3D net is the lonsdalite (or wurzite) net (Figure 8.6). Each edge of the 2D pattern now represents a chain of edges in the form of a *crankshaft*.

In Figure 8.12a the edges marked by thicker lines represent *zigzag* chains running perpendicular to the plane of the figure. The resulting 3D net is the zeolite framework CAN; the zigzags are readily apparent in Figure 8.12b.

Similarly, in Figure 8.13, Wells' 3-connected net (8, 3)-c is obtained from a 2D (2, 3)-connected pattern by the insertion of zigzag chains.

Zigzag chains are readily apparent in the structure of ice XII shown in Figure 8.10. The view along the tetragonal axis is the tiling pattern of pentagons that we showed in Figure 2.4. In this case, however, simply indicating the zigzags on the 2D pattern as in the previous examples, is inadequate. To unambiguously indicate

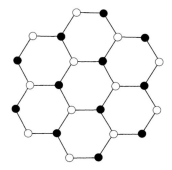

Figure 8.11 The 2D representation of the 3D net of Figure 8.6.

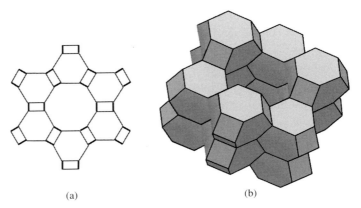

(a)   (b)

Figure 8.12 (a) The 2D representation of the zeolite framework CAN and (b) the corresponding 3D net (note the zigzag chains).

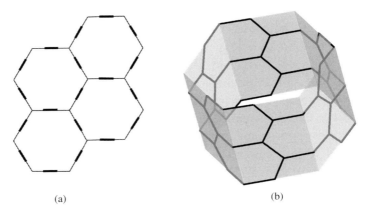

(a)   (b)

Figure 8.13 (a) The 2D representation of Wells' net (8, 3)-c and (b) a hexagonal unit cell of (8, 3)-c (The faces of the hexagonal prism are all mirror planes).

the 3D structure on the 2D diagram, the z-components of the vertices need to be specified. This has been done in Figure 8.14. There are in fact no 5-rings in the net; the rings are all heptagons and octagons.

A *saw-chain* is a sequence of edges formed from the repetition of the sequence of three edges; 'a zig, a zag and a vertical edge'. A simple example is shown in Figure 8.15. The saw-chains have been represented in the 2D representation (Figure 8.15a) by arrowed edges, and are readily apparent in the resulting 3D net (Figure 8.15b). The net is the binodal zeolite framework OFF, which consists of linked towers of cancrinite units and hexagonal prisms. Figure 8.16 represents the structure of zeolite framework FER, which exhibits a combination of crankshafts and saw chains.

These few example of relations between 2D tiling patterns and certain kinds of 3D nets should suffice to give some indication of the range of possibilities, and the enormity of the problem of systematic exploration and classification. Han & Smith

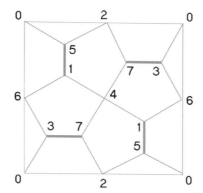

Figure 8.14 A unit cell of the ice XII net. Numbers are heights of the vertices in units of $c/8$.

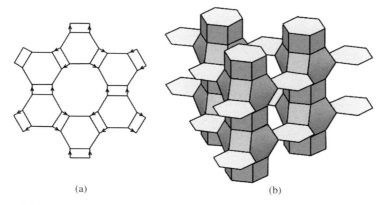

(a)   (b)

Figure 8.15 (a) The 2D tiling with saw-chains indicated by arrowed edges and (b) the resulting 3D net – the zeolite framework OFF.

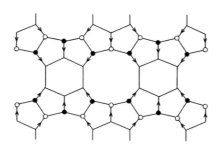

Figure 8.16 Zeolite framework FER.

(1999) have made an extensive investigation of 3D nets derivable from 2D tilings by the three methods outlined above. Their work listed hundreds of 3D nets together with detailed topological information (ring structure, polyhedral subunits, etc.), and identified those that correspond to known zeolite frameworks and other

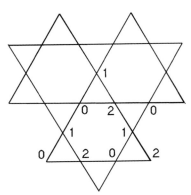

Figure 8.17 The β-quartz net derived from the kagome pattern.

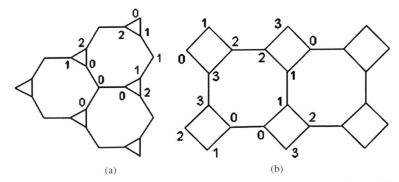

(a)   (b)

Figure 8.18 The (10, 3)-a net represented in 2D: (a) View along a threefold axis. Numbers denote heights of vertices in units of $c/3$. (b) View along a fourfold axis; heights in $c/4$ units.

crystalline structures. Their starting point was the data on 2D nets in the catalogue of the Consortium of Theoretical Frameworks (CTF), which maintains databases at the University of Chicago.

Another method of deriving 3D nets from 2D tilings is by interpreting a tile as a polygonal helix. The net for β-quartz (Figure 7.1), for example, is obtained in this way from the 'kagome' pattern (Kepler's 3.6.3.6) as in Figure 8.17, where the numbers denote heights of the vertices in units of $c/3$. Figure 8.18 is a similar representation of (10, 3)-a, viewed along the three- and fourfold axes.

8.7 Disclination Networks

The 'perfect' polytetrahedral structure is the polytope {3, 3, 5}, in which regular tetrahedra are packed, five around every edge and twenty around every vertex. The structure can occur only in the curved space S_3. Five tetrahedra in E_3, sharing an edge and leaving no gaps between the faces containing the common edge, must deviate from regularity. The required deformations becomes extreme if one tries to tile E_3 by tetrahedra with five round *every* edge. This is because the structure 'wants' to fold into E_4 and produce a $\{3, 3, 5\}$. In a polytetrahedral structure in E_3, edges shared by *six* tetrahedra are necessary to compensate.

Sadoc and Mosseri (1999) have considered the introduction of defects into $\{3, 3, 5\}$, that reduce the curvature. In the limit one obtains a polytetrahedral structure in *flat* space, E_3. The defects in the standard Frank–Kasper phases are lines of *negative* disclination, which are comprised of edges shared by *six* tetrahedra instead of five. They are necessary to counteract the curvature of the $\{3, 3, 5\}$ that would result if all edges were shared by five tetrahedra. Disclination lines in polytetrahedral structures cannot have end points. They form disclination networks. Some disclination networks are simple and obvious: for β-W for example the vertices of the disclination network are 2-connected (corresponding to the 14-coordinated atoms). The network simply consists of three orthogonal families of non-intersecting lines, passing through the hexagons of Figure 3.23. For the Friauf-Laves phase (Figure 3.17) the disclination network is the D net. In some Frank-Kasper phases the disclination network is much more complicated.

A more severe distortion is involved in 'non-standard' Frank–Kasper phases, when *four* tetrahedra share faces around a common edge. These are lines of *positive* disclination. The non-standard coordination polyhedra that can result when positive disclinations are present are discussed in Appendix A6 of Sadoc and Mosseri. The numbers N_4 and N_5 of positive and negative disclinations in a coordination shell must satisfy $2N_4 + N_5 = 12$ (according to the dual of the $(c = 3)$ formula of Section 2.11). Coordination number 13 ($N_4 = 1, N_5 = 10$ and $F_6 = 2$) occurs in γ-brass. The coordination shells of the 15-coordinated atoms in γ-brass are also of non-standard type ($N_4 = 2, N_5 = 8$ and $N_6 = 5$). Positive disclination also occurs in α-Mn (described in Section 6.5); the coordination polyhedra of the 14-coordinated atoms in α-Mn are of non-standard Frank-Kasper type.

8.8 Topological Classification of Tilings

Space-fillings by saddle polyhedra were alluded to in Chapter 3. We take up the topic again here because the main interest in these 3D tilings is motivated by interest in the nets constituted by their vertices and edges. The faces of the polyhedral tiles are rings (minimal circuits) of the nets.

Ferro & Fortes (1985) considered the question that Fedorov had asked (and answered), but generalised it by dropping the requirement that the polyhedra be convex. They considered non-convex polyhedra (with skew polygonal faces) that tile E_3, with the tiles related to each other by translations. They presented a general procedure to generate Schlegel diagrams (2D graphs representing the topology of edges and vertices of polyhedra) for an infinite family of 3D tiles, starting from the

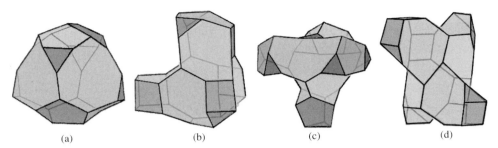

Figure 8.19 Some space-filling parallelohedra (O'Keeffe 1999). The symbol following the name of a parallelohedron lists the polygonal faces: (a) K16 $\{3^4.5^8.8^4\}$, (b) K18 $\{3^4.4^6.6^2.7^4.10^2\}$, (c) K20-1 $\{3^8.5^6.9^6\}$, (d) K20-3 $\{3^4.4^8.7^4.8^2.10^2\}$.

14-faces truncated octahedron (the only Fedorov polyhedron giving rise to a tiling with a 4-connected net). Schlegel diagrams were presented for space-filling polyhedra with 16, 18, 20 and 26 faces. Some of these had been discovered much earlier, by applying topological transformations to the space-filling by truncated octahedra (Figure 3.2) (Smith 1953).

O'Keeffe (1999) demonstrated that the polyhedra of Ferro & Fortes are realisable in E_3 with all edge lengths equal – so that the associated nets are 4-connected nets representing packings of equal spheres – and discovered several more. A few of them are shown in Figure 8.19.

Topological invariants (vertex symbols, coordination sequences, etc.) encode information about the topological properties of nets. A *Delaney symbol* (Dress 1983, 1987) is a topological invariant that encodes *complete* topological information about any tiling (in E_n, S_n or H_n) and, hence, about the net associated with the tiling.

The Delaney symbol for any given tiling is a symbol string constructed with reference to a triangulation of the tiles. Writing it down is a fairly straightforward procedure that we shall outline below. What is not so straightforward is the inverse problem – the construction of the tiling or its net from a given Delaney symbol. In a remarkable paper in *Nature* (Delgado-Friedrichs *et al.* 1999a) the success of a project to solve this problem was reported. Its solution draws upon sophisticated methods and results from group theory, topology, combinatorics and computer search algorithms. The authors call the Delaney symbol an 'inorganic gene'. They have applied their methods to enumerate and classify various types of uninodal, binodal and trinodal 4-connected nets in E_3.

The Delaney symbol for a 2D tiling is obtained by the following procedure. Place a point on every edge and inside every tile. Vertices, edge points and tile-points are designated 0, 1 and 2, respectively. One can then, in an obvious way, produce a subdivision of the tiles into triangles 012. Label the symmetrically distinct kinds of triangle A, B, C, etc. For a triangle of type A, identify the types of triangles that share with it the edge opposite its vertex 0 (i.e. the edge 12), the edge opposite its vertex

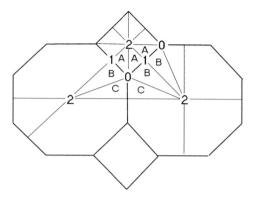

Figure 8.20 Triangulation of a tiling pattern of squares and octagons for construction of its Delaney symbol.

1 and the edge opposite its vertex 2. This gives an ordered set of three letters. This is followed by two numbers m_0 and m_1. $2\,m_0$ is the number of triangles surrounding the vertex opposite the edge 01 (i.e. vertex 2) of the triangle A and $2\,m_1$ is the number of triangles surrounding the vertex opposite the edge 12. (In other words, the vertex figure and the original tile, that the triangle A belongs to, are respectively an m_0-gon and an m_1-gon.) The procedure is repeated for a triangle of type B, for a triangle of type C, and so on. For 2D the description of the procedure, given above, is unnecessarily clumsy; we have described it in a manner that readily generalises to 3D and to higher dimensional cases. Figure 8.20 illustrates a simple example (Kepler's tiling 4.8^2). Following the procedure produces the 'gene' AAB43 BCA83 CBC83.

For a 3D tiling, vertices, edge points, face points and cell points are labelled 0, 1, 2 and 3. The tiling is then subdivided into tetrahedra. For each tetrahedron type, the Delaney symbol lists, in order, the tetrahedron type it shares with the face opposite its type 0 vertex, which it shares with the face opposite its type 1 vertex, and so on. Thus, we have a string of four letters for each tetrahedron type. This string is followed by three numbers m_0, m_1 and m_2, where m_0 is the number of tetrahedra around the edge opposite 01 (i.e. edge 23), m_1 is the number around the edge opposite the edge 12 (i.e. 30) and m_2 is the number around the edge opposite 23 (i.e. 01). As an example, the Delaney symbol for the polyhedron packing Fd $\overline{3}$m: $3^3 + 3.6^2$ is AAAB334 BBCA334 CCBC634. The generalisation to higher dimensions should now be clear. The symbol for the regular polytope $\{p, q, r\}$ (a tiling of S_3), for example, would be AAAAApqr. For a particular tiling the symbol string, or 'gene', is, of course, unique apart from the trivial changes brought about by permuting the letters A, B, …, or by replacing them by different labels.

Of course, not every symbol string that looks like a Delaney symbol is legitimate. It must satisfy certain combinatorial criteria, some of which are obvious (e.g. if A shares an edge (or a face) with B then B shares an edge (or a face) with A), others less

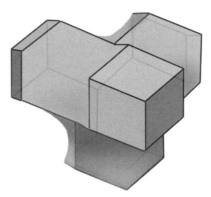

Figure 8.21 A parallelohedron K18-2 $\{4^{12}.8^4\}$ with tetrahedral symmetry: the tile for generating the net 3–175 of Delgado-Friedrichs *et al.* (1999a).

obvious. Delgado-Friedrichs *et al.* have solved the problem of deciding whether a given legitimate Delaney symbol for a 3D tiling refers to a tiling in E_3, in S_3 or in H_3, developed algorithms for enumerating particular types of tilings in E_3. They applied the methods particularly to tilings in E_3 with uninodal, binodal and trinodal 4-connected nets. Enumeration of the nets requires a way of identifying a net when it occurs as the net associated with more than one tiling (which can happen because not every ring of the net is a face of the tiling). Coordination sequences were employed to avoid this double counting, and also for recognising known nets such as zeolite frameworks. Finally, optimisation methods were used to construct actual tilings.

A 4-connected vertex of a tiling can be designated as *simple* or *quasi-simple* according to whether its vertex figure (a tiling of S_2) is or is not a tetrahedron. (e.g. the tiling associated with the diamond net, illustrated in Figure 8.5, is quasi-simple because every pair of edges with a common vertex belongs to *two* hexagonal faces; the vertex figures therefore have 'digon' faces: $V = 4$, $E = 12$, $F_2 = 6$, $F_3 = 4$). In E_3 there are found to be 9 simple and 285 quasi-simple uninodal 4-connected tilings, giving rise to (at least) 154 topologically distinct nets. There are 117 binodal and 926 trinodal simple, 4-connected tilings.

Six illustrations of unusual tilings were included in the *Nature* paper. More are to be seen on Delgado-Friedrichs' web site. They have a strange beauty. Net 3-175 with symmetry P $\overline{4}$3m is constructed from a single *parallelohedron* (Figure 8.21). (Incidentally, it is a counterexample to two questions raised by O'Keeffe (1999): whether all such parallelohedra necessarily have triangular faces, and whether the examples found by Ferro & Fortes and O'Keeffe exhaust the possibilities for parallelohedra with 16, 18 or 20 faces.)

An interesting related question concerns the 4-connected nets that can be formed from the edges of a *monohedral* tiling – the problem is to enumerate and classify the polyhedral tiles with a given number of faces and three edges at each vertex that can be packed to fill space. Delgado-Friedrichs and O'Keeffe (2005) have shown that there are 23 such tiles with 14 faces (the minimum number of faces possible for a tile with the required property). The simplest is the 'Kelvin polyhedron' (truncated octahedron). There are 136 with 15 faces and 710 with 16 faces.

Further technical details about tilings and Delaney symbols are to be found in papers by Dress *et al.* (1993), by Delgado-Friedrichs & Huson (1998), by Delgado-Friedrichs *et al.* (1999a, b) and on Daniel Huson's web site.

8.9 Interwoven Nets

A well-known example of nets occupying the same 3D space, where the edges of each net pass through rings of the other, is the double-diamond structure (Figure 8.22). The two nets are related to each other by a translation (½, ½, ½). Important examples of pairs of interwoven nets are the labyrinth graphs of the triply periodic surfaces that we shall meet in Chapter 9. For a *balance* surface the two labyrinth graphs are congruent, like the two nets in Figure 8.22, which are the labyrinth graphs of the 'D surface'.

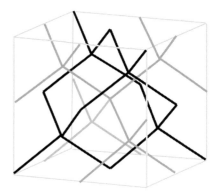

Figure 8.22 Interwoven D nets.

Wells (1977) discovered quite a few interesting examples where two or more congruent nets are *interpenetrating* or *interwoven*.

The examples we illustrate here were derived by Hyde & Ramsden (2000b) by mapping tree structures on to minimal surfaces (see also Hyde & Ramsden 2003 for further explorations of the interesting relation between 3D nets and 2D hyperbolic space). Figure 8.23 shows a hexagonal unit cell for a configuration of *three* interwoven 3-connected nets – each net is Wells' (8, 3)-c. Hyde and Ramsden derived it from the topological equivalent drawn on Schwarz's H surface (Schwarz 1890). Families of four and of eight interwoven copies of (10, 3)-a were also found by Hyde and Ramsden. The simplest way of visualising them seems to be to imagine them drawn on the surfaces of the Coxeter–Petrie polyhedra {4, 6}, {6, 4} and {6, 6}, as shown in Figure 8.24.

Figure 8.23 A unit cell for a configuration of 3 interwoven (8, 3)-c nets.

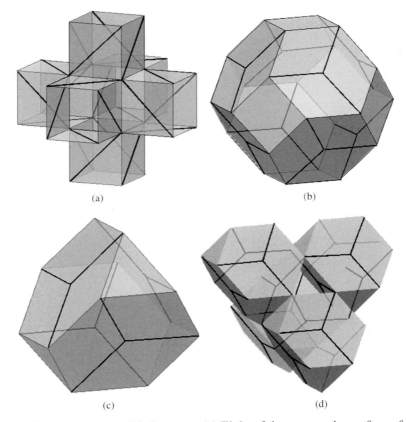

Figure 8.24 Interwoven (10, 3)-a nets. (a) Eight of the nets on the surface of the regular polyhedron {4, 6}. All vertices of the polyhedron are also nodes of the nets. (b). The same configuration of eight (10, 3)-a nets inscribed on the polyhedron {6, 4}. Polyhedron vertices are in this case mid points of net edges. (c) and (d) Four (10, 3)-a nets inscribed on {6, 6}.

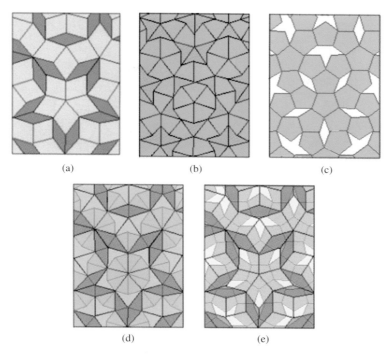

(a) (b) (c)

(d) (e)

Plate I Penrose tiling patterns. (a) Rhomb pattern; (b) kite and dart; (c) pentagon pattern. (d) Superimposition of (a) and (b) – observe that all the 'fat rhombs' are decorated identically by the kite and dart pattern, as are all the thin rhombs. (e) The superposition of (a) and (c), demonstrating the equivalence of the rhomb and the pentagon patterns.

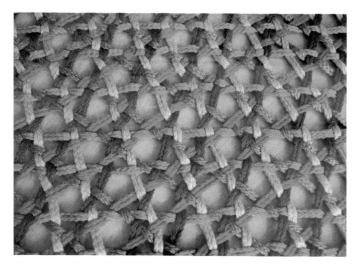

Plate II Five-stranded weaving pattern based on the Penrose rhomb tiling. Detail from a coffee table designed and made by Robert Mackay.

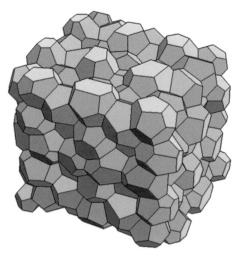

Plate III The Voronoi regions around the 152 atoms in a unit cell of $Mg_{32}(Al, Zn)_{49}$.

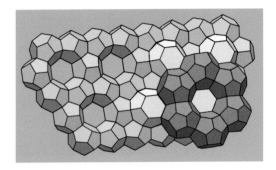

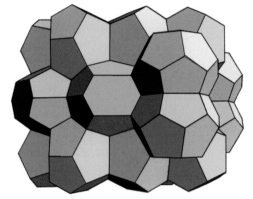

Plate IV Two views of the dissection of space into Voronoi regions of Zr_4Al_3.

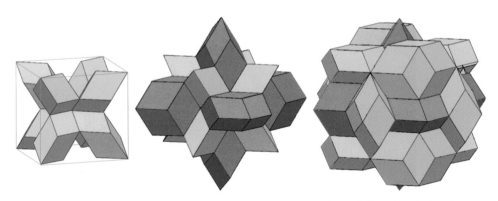

Plate V A periodic packing of prolate Kowalewski units and rhombic dodecahedra; (a) the cubic unit cell contains eight prolate units; (b) the rhombic dodecahedra are centered at mid-ponts of faces of the unit cell and (c) at mid points of the edges of the unit cell.

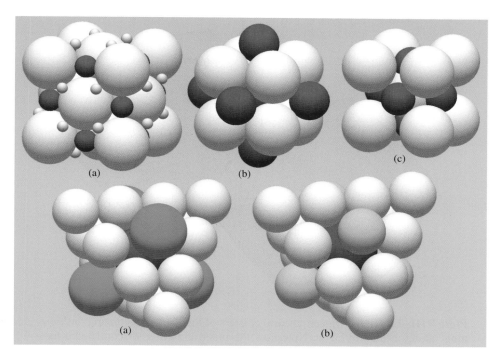

Plate VI Top row: (a) the fcc sphere packing with smaller spheres in the octahedral interstices (red) and in the tetrahedral interstices (green); (b) spheres occupying the voids in a primitive cubic array of close packed spheres; (c) a bcc array of spheres with smaller spheres occupying the voids. Bottom row: (a) a portion of the Laves type packing of spheres of three kinds, with optimised packing fraction. (b) The modified AB_5 Laves phase in which half the large atoms are replaced by the small atoms.

Plate VII A tile produced from four dragon curves.

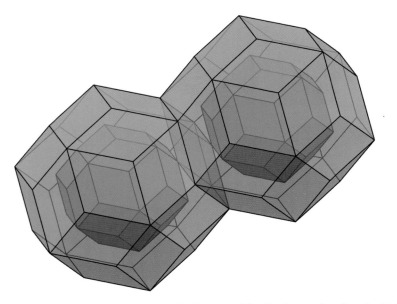

Plate VIII Two of the triacontahedral clusters of the R-phase, showing the'twin-ning' along a threefold axis. Note how the inner and outer triacontahedra have common vertices.

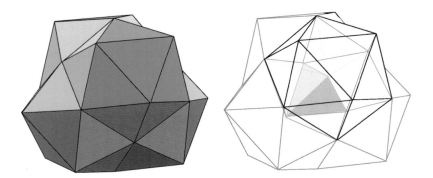

Plate IX The 26-atom γ-brass cluster as a cluster of four interpenetrating icosahedra.

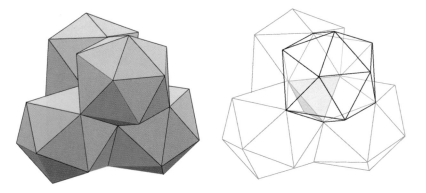

Plate X The Pearce cluster. Four icosahedra on the faces of a tetrahedron.

Plate XI A model of a portion of the infinite 'regular' polyhedron {3, 7}.

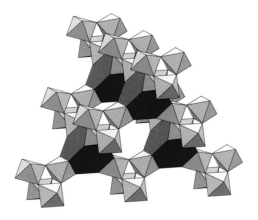

Plate XII A D-net in which nodes are alternately pyrochlore units and truncated tetrahedra.

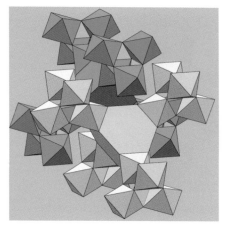

Plate XIII Two interwoven polyhedral D-nets.

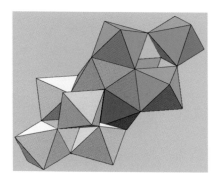

Plate XIV An icosahedron linking two pyrochlore units.

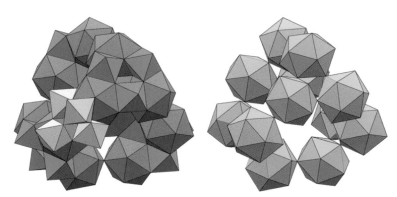

Plate XV (a) The Mg$_3$Cr$_2$Al$_{18}$ structure built from pyrochlore units and icosahedra. (b) The icosahedra form a network of vertex-sharing *L*-units.

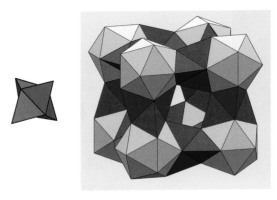

Plate XVI A *stella quadrangula* (four tetrahedra on the faces of a central tetrahedron, and (right) a polynet of icosahedra and *stellae quadrangulae.*

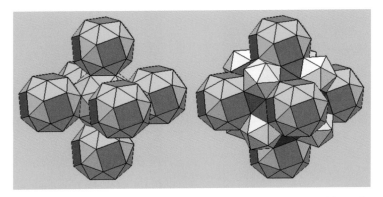

Plate XVII (a) An array of snub cubes. (b) The space filling packing of tetrahedra, icosahedra and snub cubes – a model of the structure of $NaZn_{13}$.

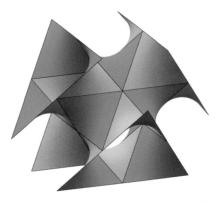

Plate XVIII A portion of the D-surface consisting of 18 Schwarz quadrilaterals.

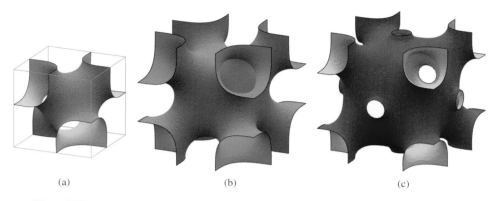

Plate XIX Three TPMS generated by reflections, discovered by Alan Schoen: (a) 1/8 unit cell of FRD; (b) a unit cell of IWP. (c) OCTO (The faces of the bounding cubes in all three cases are mirror planes.)

Plate XX A unit cell of the gyroid, discovered by Schoen.

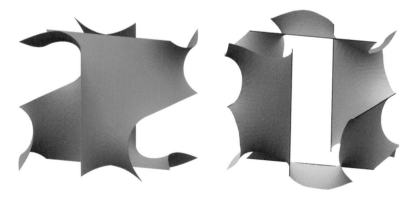

Plate XXI (a) Two nonagon generating patches in 1/8 unit cell of the surface $C(\pm Y)$, and (b) two nonagons spanned by a 'catenoid-like' generating patch of $\pm Y$.

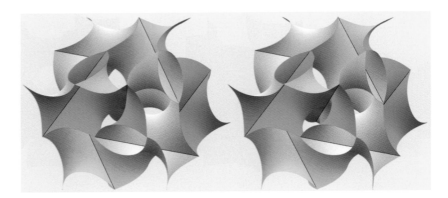

Plate XXII A stereo view of a unit cell of Elser's Archimedean screw.

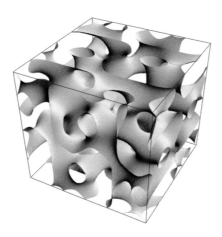

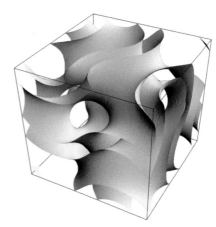

Plate XXIII von Schnering and Nesper's C(Y**), formula $3(\sin X \cos Y + \sin Y \cos Z + \sin Z \cos X) + 2(\sin 3X \cos Y + \sin 3Y \cos Z + \sin 3Z \cos X - \sin X \cos 3Y - \sin Y \cos 3Z - \sin Z \cos 3X) = 0$. This nodal surface has the same symmetries as the gyroid, I43d-I4₁32. Whether a minimal surface exists with this topology and these symmetries is not known.

Plate XXIV A 'double gyroid'. A nodal surface with the symmetry (I4₁32) and topology of the family of constant mean surfaces to which the gyroid belongs. The formula is $0.8(\sin 2X \sin Y \cos Z + \sin 2Y \sin Z \cos X + \sin 2Z \sin X \cos Y) - 0.2(\cos 2X \cos 2Y + \cos 2Y \cos 2Z + \cos 2Z \cos 2X) + 0.33 = 0$.

8.10 Truncation

A straightforward way in which a net can be related to a more complicated net is by an obvious generalisation of the idea of *truncation* of a polyhedron. Truncation of a 3-connected vertex produces a triangle in the augmented net; 4-connected vertices give rise to a quadrilateral or a tetrahedron. *N*-gonal rings of the original net become 2*N*-gons in the net derived from it by truncation. The principle was employed by Heesch and Laves (1933) in their search for low-density sphere packings (see Section 4.4). We present a few further examples.

Figure 8.25 illustrates the 4-connected NbO net (so called because its vertices correspond to the positions of the atoms in NbO; the two kinds of atom alternate). It is unusual among 4-connected nets in that the four edges at a vertex have a coplanar rather than a tetrahedral configuration. Truncation produces a 3-connected net with square and dodecagon rings. The binodal 4-connected PtS net has both flat and tetrahedral nodes. Its truncation is a (3, 4)-connected net (Figure 8.26). Figure 8.27 illustrates the truncation of the binodal (3, 4)-connected net corresponding to the

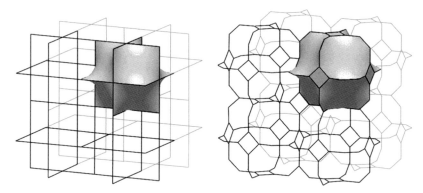

Figure 8.25 The NbO net and the truncated NbO net.

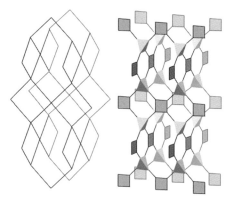

Figure 8.26 The PtS net and its truncation.

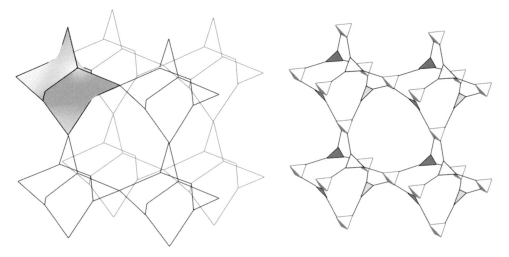

Figure 8.27 The (3, 4) boracil net and its truncation.

structure of boracite (vertices are occupied by boron atoms; edges of the net represent their linkage through the bivalent oxygen atoms, as in the representation of silicates and zeolites by their 'frameworks'). The importance for materials science of the relationship between a net and its truncation lies in the fact that complicated structures often turn out to be describable in terms of truncated versions of more commonly occurring nets. Plausible hypothetical structures leading to the design and synthesis of new materials have arisen from this aspect of nets. A fascinating

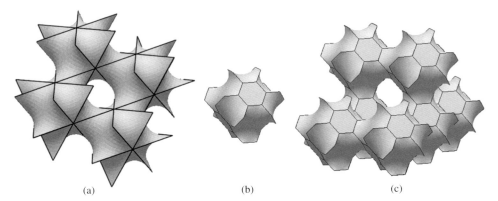

(a)   (b)   (c)

Figure 8.28 (a) A 6-connected net consisting of the edges of a 3D tiling by a saddle polyhedron whose faces are the Petrie polygons of a tetrahedron and an octahedron. (b) The saddle polyhedron obtained from this by truncation consists of edges of a tiling by a polyhedron 6.8.12. (c) The packing of the truncated polyhedra gives the 3-connected net 6.8^2 described by Wells (1977).

review of this topic has been provided by O'Keeffe *et al.* (2000), from which the three examples given above have been taken.

As a final example, consider the uninodal 6-connected net formed by the edges of the tiling by the saddle polyhedron illustrated in Figure 9.23d. The rings are skew hexagons and skew quadrilaterals (Petrie polygons of the octahedron and the cube, respectively). The six edges at a vertex are coplanar, so that truncation of the net produces a 3-connected net with 12-rings, 8-rings and flat 6-rings. This is Wells' net 6.8^2 (Figure 8.28).

8.11 Polyhedral Nets and Labyrinth Graphs

Many complex crystalline materials can be understood in terms of packings of polyhedra and the nets associated with these packings (Brunner 1981; Senechal & Fleck 1998; O'Keeffe 1998; etc.). In Figure 8.2b the zeolite framework FAU is represented by the edges of an infinite polyhedron. The polyhedron has the form of a labyrinth in which cages (truncated octahedra) are interconnected by tunnels (the hexagonal prisms). It can be characterised topologically by a *labyrinth graph* – a net whose nodes are at the centres of the cages and whose edges pass through along the tunnels. The net is in fact the diamond net (D net). Labyrinth graphs are important topological descriptors of the surfaces we shall meet in Chapter 9.

We mean by a *polyhedral net*, or *polynet* a structure in which polyhedra are centred at the nodes of a net and connected to each other either by face-sharing or by forming tunnels along the net edges. Several examples of polyhedral nets have been encountered in this and earlier chapters. For example, $MoAl_{12}$ in Figure 6.1 and αMn in Figure 6.20 are represented as polyhedral nets. In both cases the labyrinth graph is the 8-connected body centered cubic (bcc) net. In the case of $MoAl_{12}$ the nodes are icosahedra, the edges octahedra; for αMn nodes and edges are truncated tetrahedra. Figure 8.2a (zeolite framework LTA) provides a further example. The labyrinth graph for the structure is 6-connected. Truncated octahedral nodes are connected by tunnels in the form of cubes.

The labyrinth graph for both the Coxeter–Petrie infinite regular polyhedra (Figure 8.1) {4, 6} and {6, 4} is the 6-connected net of a primitive cubic lattice (P net); that for {6, 6} is the diamond net (D net). Note that the tunnels and nodes of {4, 6} are all cubes, whereas {6, 4} has no tunnel polyhedra. Since the rings of the D net are all even (all hexagons) the nodes may be of two types, that alternate (like the atoms in zinc blende ZnS). The polyhedral net {6, 6} is of this kind; the nodal polyhedra are tetrahedra and truncated tetrahedra – there are no tunnel polyhedra.

Wells (1977) employed polyhedral nets in the construction of infinite 'regular' polyhedra. Recall that the regular polyhedra of Coxeter and Petrie (Figure 8.1) have all faces regular and congruent and all vertex figures regular and congruent – the vertex figures are, however, not flat. In Wells' extended definition of a 'regular' polyhedron

Plate XI A model of a portion of the infinite 'regular' polyhedron {3, 7}.

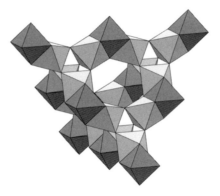

Figure 8.29 A D-net of octahedra.

the vertex figures are not required to be regular – they are, however, required to be congruent. An interesting variety of such 'regular' polyhedra were found by Wells. Plate XI, for example, can be regarded as a portion the infinite regular polyhedron {3, 7}. Its labyrinth is a D-net; icosahedra are centred at the nodes and octahedra form the tunnels. A large number of structures can be obtained with the D net as labyrinth and with regular or semi-regular polyhedra as cages and tunnels. In the simple example shown in Figure 8.29, which occurs in pyrochlore-related structures, the nodal polyhedra as well as the tunnels connecting them, are octahedra.

8.12 Polynets Containing Pyrochlore Units

The structure of the mineral pyrochlore can be described in terms of two interwoven polyhedral D nets. One of them is the polyhedral network of octahedra (Figure 8.29). The complementary polyhedral D net is the framework for the configuration

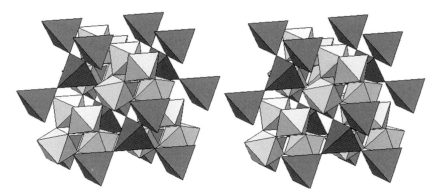

Figure 8.30 A D-net of octahedra and a complementary D-net of tetrahedra. Stereo images illustrating the pyrochlore structure.

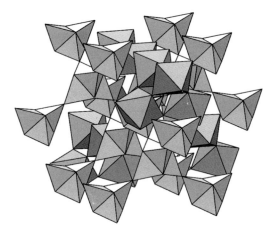

Figure 8.31 The structure of W_3Fe_2C and related materials. The edge length of the *stellae quadrangulae* is four fifths that of the octahedra.

of vertex-sharing tetrahedra shown in Figure 8.7. The resulting structure is represented by Figure 8.30. The octahedron vertices are occupied by O or F atoms. The edge octahedra are centred by Nb or Ti; the node octahedra are empty; the tetrahedron vertices are occupied by Na or Ca, and the tetrahedra are centred by O or F atoms. Nyman *et al.* (1978) have shown that a variety of crystalline materials have a similar structure. In W_3Fe_3C, for example, the octahedral net consists of W octahedra, centred by C atoms, and the tetrahedra of the interwoven structure are replaced by *stellae quadrangulae* with Fe atoms occupying their vertices (Figure 8.31).

A single material structure may be capable of two or more equally valid geometrical descriptions. A surprising example of this is that the generalised pyrochlore structure discussed by Nyman & Andersson and illustrated in Figure 8.31 can be

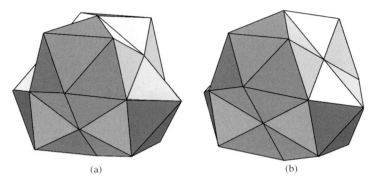

(a)   (b)

Figure 8.32 The γ-unit in two metrically slightly different versions: (a) as it occurs in γ-brass and (b) as it occurs in the pyrochlore related structures of Nyman and Andersson. In this version the skew hexagons are inversion symmetric.

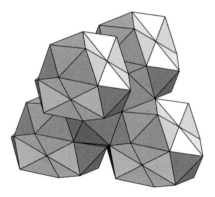

Figure 8.33 A packing of five γ-units, leading to a D-type polynet.

interpreted as a D-net of γ-brass type clusters. Figure 8.32 shows again the polyhedron corresponding to the 26-atom γ-brass cluster described in Section 6.4. If the edge length of the central *stella quadrangula* is somewhat smaller, then the skew hexagons indicated in white can be inversion symmetric, allowing these 'γ-units' to pack together to form a polyhedral D-net (Figure 8.33). The resulting structure is *identical* to that in Figure 8.31, the complementary polynet is the D-net of octahedra. We have a space-filling of γ-units and octahedra, as indicated in Figure 8.34. This structure occurs in a large number of complex alloys 'of Ti_2Ni type'. In Ti_2Ni the nickel atoms are at the vertices of the inner tetrahedra of the *stella quadrangulae*, the rest are titanium. There are 96 atoms per cubic unit cell.

A fascinating structure that can be described in terms of polyhedral nets is that of $Mg_3Cr_2Al_{18}$, described by Nyman & Andersson (1978). It provides another striking example of the way in which very complex structures can often be given

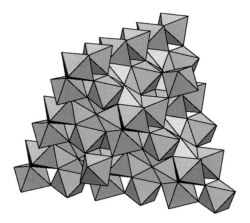

Figure 8.34 A packing of γ-units and pyrochlore units.

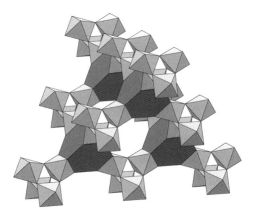

Plate XII A D-net in which nodes are alternately pyrochlore units and truncated tetrahedra.

alternative, quite different descriptions that complement each other. The voids around the polyhedral net in Plate XII are in fact large enough to accommodate another polyhedral net, congruent to the first. Thus two of these polyhedral nets can be interwoven without touching, as in Plate XIII. In the MgCrAl phase described by Nyman and Andersson voids in this structure are in the shape of icosahedra (as in Plate XIV). These icosahedra are surprisingly close to being regular.

Now, recalling that the tetrahedral cluster of four vertex-sharing icosahedra, called the *L*-unit by Kreiner & Franzen (1995), can be interpreted as a cluster of five octahedra and four icosahedra (as discussed in Section 6.7 and illustrated in Figure 6.24), we see that the MgCrAl structure can be interpreted as a D-net of Kreiner–Franzen *L*-units linked by their vertices like the tetrahedra in a Laves phase (Plate XV).

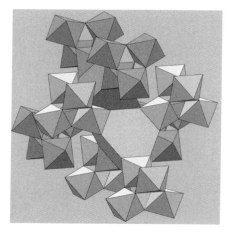

Plate XIII Two interwoven polyhedral D-nets.

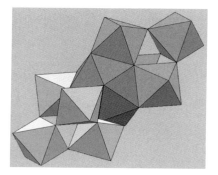

Plate XIV An icosahedron linking two pyrochlore units.

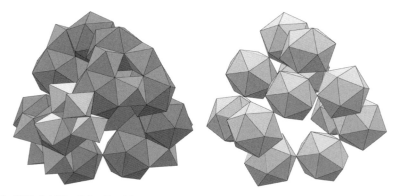

Plate XV (a) The $Mg_3Cr_2Al_{18}$ structure built from pyrochlore units and icosahedra. (b) The icosahedra form a network of vertex-sharing L-units.

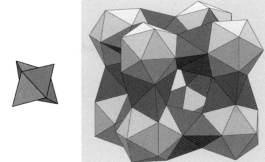

Plate XVI A *stella quadrangula* (four tetrahedra) on the faces of a central tetrahedron, and (right) a polynet of icosahedra and *stellae quadrangulae.*

A variety of intermetallics have framework-type structures in which *stellae quadrangulae* are basic building blocks (Haüssermann *et al.* 1998). The *stella quadrangula* is a very versatile structural unit. A particularly fascinating example is the
$NaZn_{13}$ structure. If three tetrahedra and an icosahedron share a single edge, a very
small deviation from regularity is all that is required to close the gaps between faces.
(This fact was encountered already in our presentation of the Bergman cluster as a
polytetrahedral structure.) It follows that a *stella quadrangula* can provide a link
between two icosahedra, and a polynet can be built with icosahedra centred on a
primitive cubic lattice, linked by stella quadrangulae, as in Plate XVI. This vertices
of this polynet and the centres of its icosahedra correspond to the positions of the Zn
atoms in structure of $NaZn_{13}$ (Haüssermann *et al.* 1998). The Na atoms are located
in the large 'cages' formed by the structure. Each Na atom is coordinated to 24 Zn
atoms. The coordination shells have the form of the semi-regular polyhedron 4.3^5 –
Kepler's *snub cube*. The whole structure can be described geometrically in terms of
a space filling arrangement of tetrahedra, icosahedra and snub cubes (Plate XVII),
with an almost negligible deviation from perfect regularity.

8.13 Augmented Nets

The previous two sections have served to suggest ways in which a complex structure
can often be described and understood in terms of a relatively simple net. In such a
description, nodes and edges of the net represent, not single atoms and bonds, but
more complicated subunits. We quote from Schindler *et al.* (1999): 'vertices may be
single atoms, dimers, coordination polyhedra, clusters of atoms or clusters of coordination polyhedra; edges may be single bonds or sets of several chemical bonds...'. It
turns out that only a few of the simplest 3-, 4-, and (3, 4)-connected nets seem to be

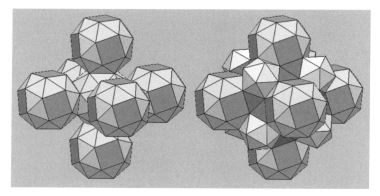

Plate XVII (a) An array of snub cubes and (b) the space-filling packing of tetra-hedra, icosahedra and snub cubes – a model of the structure of $NaZn_{13}$.

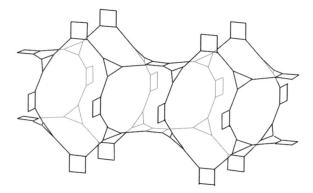

Figure 8.35 The net obtained by truncation of the Pt_3O_4 net.

needed for the description of a very large number of complicated crystalline materials that can be regarded as being built around an underlying framework or 'scaffolding'.

Identification of the simple nets that underlie the structures of complex alloys and other crystalline materials often reveals close relationships among otherwise unrelated structures.

As discussed in Chapter 6, many complex alloys consist of clusters built from nested polyhedral shells. These clusters combine by sharing of atoms to form a framework, or net, of clusters. The descriptions of γ-brass (Section 6.4) and α-Mn (Section 6.5) are striking examples.

The synthesis of a material (MOF-14) with remarkably large pores, which can thus adsorb large amounts of gases and organic solvents, has been reported (Li *et al.* 1999; Chen *et al.* 2001). The Pt_3O_4 net has 3-connected and 4-connected vertices. Its truncation gives the net shown in Figure 8.35. The structure of MOF-14 can be described in terms of an underlying framework consisting of two of these

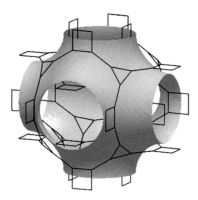

Figure 8.36 Two interwoven truncated Pt_3O_4 nets.

nets, interwoven. In Figure 8.36, we see how the two interwoven nets lie, respectively, within the two labyrinths of the P surface. In the actual material, each vertex corresponds to molecular units rather than single atoms. The structure has considerable stability on account of the interweaving. The porosity of zeolites is responsible for their practical importance in industry. The possibility of actually designing and synthesising stable porous materials with very much larger pores is expected to lead to many important applications (Ferey 2001; Davis 2002).

The approach to understanding and describing complex structure, based on nets formed from clusters, has been extensively developed by Andersson (1978), Nyman & Andersson (1978; 1979), Hellner & Koch (1981), Chabot *et al.* (1981) and Nyman & Hyde (1981).

9

Triply Periodic Surfaces

9.1 Minimal Surfaces

In this chapter we shall be discussing continuous smooth surfaces with space group symmetries. Much of the literature on *triply periodic* surfaces deals principally with *minimal surfaces* (TPMS). Minimal surfaces occur in natural structures in a wide variety of contexts. The TPMS have been observed in the form of interfaces, for example in lipid bilayers and diblock copolymers (Mackay 1990; 1993; Gozdz & Holyst 1996). Moreover, interesting geometrical relationships exist between atomic arrangements in crystalline substances, and TPMS (von Schnering & Nesper 1987; Hyde *et al.* 1996; Klinowski *et al.* 1996; Ozin 1999; Hyde & Ramsden 2000a, b; Hyde & Ramsden 2003).

This section is in the nature of a preamble; it will not be concerned with periodicity. Its purpose is to introduce briefly a few of the key properties of minimal surfaces.

The mathematical theory of minimal surfaces lies outside the scope of this present work; a nice introduction to the subject is Osserman (1986) and a thorough presentation of this complicated topic has been provided by Nitsche (1989). The problem of determining mathematically the shape of the surface patch of least area with a given closed curve as boundary (e.g. a soap film spanning a wire loop) is known as 'Plateau's problem' (Plateau 1873; Almgren 2001; Brakke 2001). Readers interested in the underlying differential geometry may consult, e.g., Coxeter (1969), (1989), Lord & Wilson (1984), Hyde *et al.* (1996), etc. The important point to be noted here is that *minimal surfaces are surfaces with zero mean curvature*. The Gaussian and mean curvatures at a point on a surface in E_3 are $K = k_1 k_2$ and $H = (k_1 + k_2)/2$, where k_1 and k_2 are the two principle curvatures. Two minimal surfaces were known a century before Plateau's contribution. They are the *catenoid* and the *helicoid* (in cylindrical coordinates, $\rho = \cosh(z)$ and $z = \phi$), shown to be minimal by Laplace and Meusnier, respectively.

Weierstrass (1866) found a surprising correspondence between minimal surfaces in E_3 and complex analytic functions. (See Hyde & Andersson (1985) or Hyde (1989) for the differential geometry associated with the Weierstrass formula.) Specifically, if $F(w)$ is an analytic function of a complex variable $w = u + iv$, then the surfaces given by the parametric forms $\mathbf{r}(u, v)$ and $\mathbf{r}'(u, v)$, where

$$\mathbf{r}(u, v) + i\mathbf{r}'(u, v) = \int_\omega F(w)\mathbf{n}(w)dw, \quad \mathbf{n}(w) = (w - w^2, i(w + w^2), 2w)$$

are minimal surfaces. The two surfaces $\mathbf{r}$ and $\mathbf{r}'$ are said to be *adjoint*. From a given minimal surface an infinite family of minimal surfaces is obtained by applying the *Bonnet transformation* (Bonnet 1853), which, in terms of the Weierstrass function F, is simply $F \to e^{i\theta}F$. For $\theta = \pi/2$ we get the adjoint surface. The whole one parameter family of surfaces provided by varying θ from 0 and $\pi/2$ has remarkable properties. All the surfaces are *metrically* identical – the continuous deformation is a *bending* of the surface. Also, the direction of the normal at each point on the surface remains fixed throughout the sequence and the neighbourhood of each point rotates around the normal through the Bonnet angle θ. The catenoid and the helicoid provide a simple example of an adjoint pair. Some of the intermediate surfaces encountered when they are transformed into each other by the Bonnet transformation are shown in Figure 9.1.

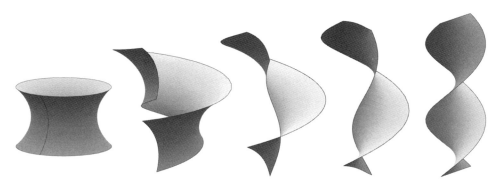

Figure 9.1 The catenoid ($\theta = 0$) and surfaces obtained from it by the Bonnet transformation: $\theta = \pi/8$, $\theta = \pi/4$, $\theta = 3\pi/8$ and the helicoid ($\theta = \pi/2$).

Two special categories of *geodesics* on a minimal surface are related to the symmetries of the surface. If part of the boundary of a minimal surface patch is a straight line then the entire minimal surface contains the whole line and is invariant under 180° rotation about the line, and if a minimal surface meets a plane at right angles then reflection in the plane is a symmetry of the surface. These two special kinds of geodesic – straight lines and planar geodesics – are interchanged, for a surface and its adjoint.

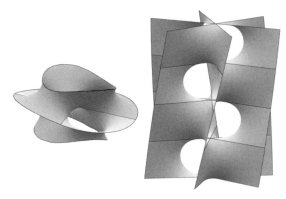

Figure 9.2 Minimal surfaces without self-intersection. (a) The central portion of Costa's surface (Costa 1982; 1984; Hoffman 1987). The surface has two diad axes. Variants exist for different numbers of embedded diad axes. It can be visualised by thinking of a catenoid intersected by a plane, and replacing the circle of intersection by 'tunnels'. (b) A minimal surface due to Karcher (1989b). It can be thought of as three helicoids intersecting along their common axis with the line of intersection replaced by a system of tunnels. A simpler minimal surface related in a similar way to a pair of intersecting planes is Scherk's singly periodic minimal surface (Scherk 1834).

In general, minimal surfaces obtained from an arbitrary Weierstrass function $F(w)$ will have self-intersections. Minimal surfaces in E_3 that can be analytically extended indefinitely without intersecting themselves are of special interest. The catenoid and helicoid belong to this category; the intermediate surfaces do not. In Figure 9.2 we show just two examples of other known minimal surfaces without self-intersection. A very large number of TPMS without self-intersection are now known. This is the main topic of the following sections.

9.2 Schwarz and Neovius

Schwarz (1890) found the Weierstrass function for a minimal patch spanning an equilateral skew quadrilateral all of whose angles are $\pi/3$ (i.e. the Petrie polygon of a regular tetrahedron), thereby solving the Plateau problem for this boundary. (Incidentally, the Weierstrass function is in fact $(1 - 14w^4 + w^8)^{-1/2}$). The surface can be extended to a triply periodic surface without self-intersections by repeatedly applying diad rotations about the straight edges. The complete surface partitions E_3 into two congruent regions (labyrinths) whose topological configuration can be described by labyrinth graphs. The pair of labyrinth graphs is the 'double-diamond' net (Figure 8.22); accordingly, the surface is called the D-surface. The portion of it shown in Plate XVIII, consisting of 18 of the Schwarz quadrilaterals, surrounds each node of the labyrinth graph.

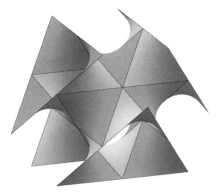

Plate XVIII A portion of the D-surface consisting of 18 Schwarz quadrilaterals.

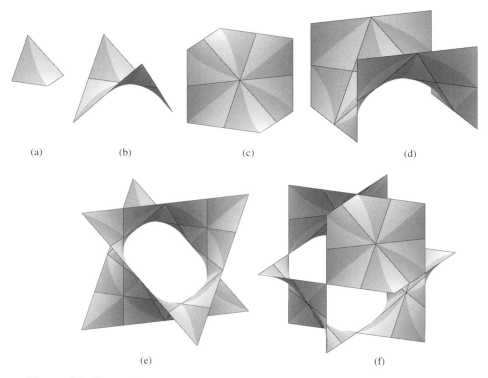

Figure 9.3 Generating patches for D: (a) the smallest generating patch, (b) the Schwarz patch spanning the Petrie polygon of a regular tetrahedron, (c) Petrie polygon of a cube, (d) an octagonal patch, (e) a 'catenoid-like patch spanning two equilateral triangles and (f) A unit cell of D.

Schwarz's quadrilateral patch is a generating patch for the D-surface. In general, a *generating patch* for a TPMS is a patch of the surface bounded by straight edges, so that the whole surface is generated by repeated diad rotations. Examples of generating patches for the D-surface are shown in Figure 9.3.

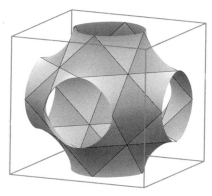

Figure 9.4 A unit cell of the P-surface. The bounding cube consists of mirror planes.

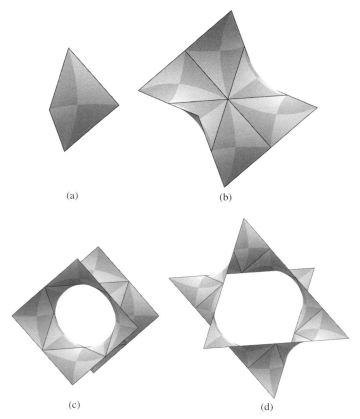

Figure 9.5 Generating patches for the P-surface: (a) the smallest generating patch, (b) an equilateral skew hexagon with all angles $\pi/3$ (Petrie polygon of a regular octahedron), (c) a square 'catenoid-like' generating patch and (d) a catenoid-like generating patch.

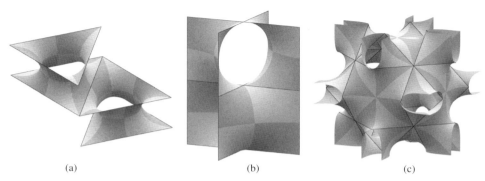

(a) (b) (c)

Figure 9.6 (a) The contents of the unit cell of Schwarz's H-surface. The generating patch is a catenoid-like patch spanning two equilateral triangles. (b) Contents of a unit cell of Schwarz's tetragonal surface CLP. (c) Unit cell of Neovius's surface C(P) (the 'complement' of P) which has the same framework of embedded straight lines as P.

Schwarz also discovered the P-surface, a unit cell of which is shown in Figure 9.4. The smallest generating patch for P is a quadrilateral with two angles $\pi/2$ and two angles $\pi/3$. Figure 9.5 shows a variety of larger generating patches.

Two other TPMS found by Schwarz, and one due to Schwarz's student Neovius (1883), are illustrated in Figure 9.6.

After the pioneering work of Schwarz and Neovius, a very long time elapsed without further advance in the study of TPMS. Stessmann (1934) applied Schwarz's method to find the Weierstrass functions for those TPMS that have rectilinear quadrilateral generating patches. There are just six of them (a result established by Schoenflies); two of them are of course D and P, the other four have rectilinear self-intersections.

The five TPMS of Schwarz and Neovius remained the only known TPMS free of self-intersections, until Alan Schoen discovered twelve more (Schoen 1970).

9.3 Schoen's Surfaces

TPMS obtained from generating patches bounded by straight lines are necessarily *balance* surfaces: the two labyrinths are congruent, since diad rotations about straight lines embedded in the surface interchange the two labyrinths (Molnár 2002). Schoen considered patches bounded by reflection planes, and thereby was able to discover TPMS that are *not* 'balanced'. These surfaces partition E_3 into two *different* labyrinths. The surfaces shown in Plate XIX are of this kind. They can be generated from a patch contained in a cube by reflections in the cube faces. In the name I-WP given by Schoen one of his surfaces the I indicates the space group

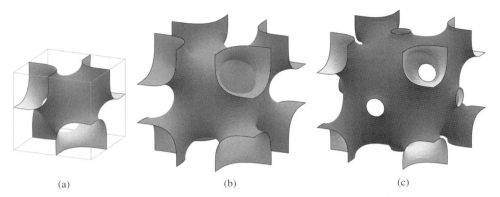

(a)   (b)   (c)

Plate XIX Three TPMS generated by reflections, discovered by Alan Schoen: (a) 1/8 unit cell of FRD, (b) a unit cell of IWP, (c) OCTO (the faces of the bounding cubes in all three cases are mirror planes).

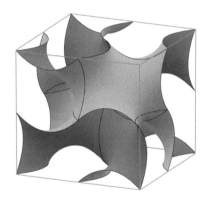

Plate XX A unit cell of the gyroid, discovered by Schoen.

symmetry (Im3̄m) and the WP stands for 'wrapped package' because of the appearance of one of the labyrinth graphs – actually the NbO net (Figure 8.25). The Weierstrass function for I-WP has been obtained by Lidin *et al.* (1990).

Schoen's most remarkable discovery is the *gyroid* (the G-surface, often referred to also as the surface Y*). Plate XX shows the appearance of the unit cell. The gyroid is a TPMS with *no embedded straight lines* that is nevertheless a *balance* surface – the two labyrinths are enantiomorphs, transformed into each other by $\bar{3}$ and $\bar{4}$ transformations with centres on the surfaces. Each labyrinth graph is a (10, 3)-a. We have already mentioned that the surfaces D and P are adjoint. Amazingly, the one-parameter family of surfaces given by the Bonnet transformations between D ($\theta = 0°$) and P ($\theta = 90°$) gives the gyroid at $\theta = 38.015...°$ (Figure 9.7).

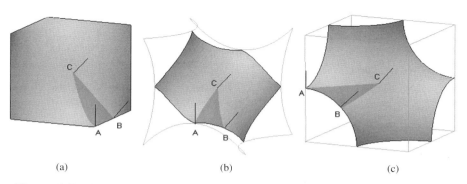

(a)   (b)   (c)

Figure 9.7 The Bonnet transformation applied to the D surface: (a) ($\theta = 0$) a hexagonal patch of the D surface, with normals indicated at the vertices of the fundamental triangle ABC, (b) ($\theta = 38.015\ldots^\circ$) a hexagonal patch on the gyroid (the larger hexagon indicated by faint lines is the portion of the surface contained in 1/8 of a unit cell) and (c) ($\theta = 90^\circ$) a hexagonal patch of P. Observe that the normals at A, B and C remain unchanged and that the surface rotates about them, through angle θ. Observe also how straight lines and curves lying in planes of reflection symmetry are interchanged between (a) and (c).

9.4 Generating Patches

In a remarkable sequence of papers (Fischer & Koch 1987; 1989a, b, c; 1996; Koch & Fischer 1988; 1989a, b) very many hitherto unknown surfaces were described in detail and illustrated by photographs of models. They were derived by investigating systematically the possibilities for generating patches that would lead to triply periodic minimal balance surfaces (TPMBS) without self-intersections. The symmetry of TPMBS can be characterised by a *pair* of space groups: the symmetry group G of the surface itself including transformations that interchange the two sides of the surface (and hence the two labyrinths), and the symmetry group of *one* labyrinth – a subgroup H of G, of index 2. The framework of straight lines embedded in the surface is the collection of diad axes that belong to G but not to H. Koch & Fischer (1988) gave an exhaustive list of all possible G–H pairs that are not incompatible with the existence of a TPMBS. There are 547 such pairs.

In particular, the G–H symmetries of the TPMBS of Schwarz and Neovius are: D: $Pn\bar{3}m$–$Fd\bar{3}m$, P and C(P): $Im\bar{3}m$–$Pm\bar{3}m$, H: $P6_3/mmc$–$P6m2$, CLP: $P4_2/mcm$–$P4_2/mcm$. (This last instance draws attention to the fact that a proper subgroup of an *infinite* group G may be isomorphic to G) The gyroid has symmetry $Ia\bar{3}d$–$I4_132$.

The method for finding new minimal surfaces then becomes a matter of studying the frameworks of diad axes to see if, and how, they can be spanned by generating patches. A generating patch may be topologically 'disc-like', as for example the polygonal patches for D in Figures 9.3a–d, those for P in Figures 9.5a, b or the hexagonal patch for CLP that forms a quarter of Figure 9.6b or 'catenoid-like', for example the patches spanning pairs of equilateral triangles for D, P or H. Other possibilities are 'branched catenoids', 'multiple catenoids (MCs)', 'catenoids with spout-like attachments' and 'infinite strips' (Figure 9.8).

Two particularly interesting TPMBS discovered by Fischer & Koch (1987) are $\pm$Y and C($\pm$Y), with symmetry Ia$\bar{3}$-Pa$\bar{3}$. There are no reflection symmetries and the embedded diad axes in these cases form three mutually orthogonal families of

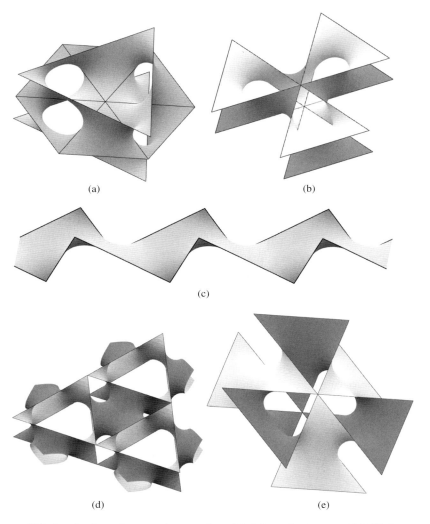

(a)   (b)

(c)

(d)   (e)

Figure 9.8 A selection of generating patches. (a) A branched catenoid BC1 (symmetry P6$_3$22–P6$_3$), (b) an MC1 (P6$_3$/mcm–P$\bar{6}_2$m), (c) an infinite strip ST1 (P6$_2$22–P6$_4$22), (d) three of the catenoids with spout-like attachments that are generating patches for Schoen's C(H) (P6$_3$/mmc–P$\bar{6}$m$_2$) and (e) a variation of the 'MC' type (Lord 1997) spanning the same layers of diad axes as the MC1 surface. The symmetry of the resulting surface is P$\bar{3}$1m-P$\bar{3}$1m (2*c*).

parallel straight lines. They have no intersections and therefore there are no closed circuits consisting entirely of straight lines. Generating patches for these two surfaces necessarily have curved as well as straight and flat edges. The curved edges can be taken to be geodesics. Fischer & Koch (1987; 1989c) presented disc-like 18-gon generating patches. Plate XXI shows an alternative description in terms of nonagons and a nonagonal catenoid. The portions of the surfaces shown in the figure are related to a cube corresponding to an eighth of a unit cell. The eight vertices of the cube are inversion centres (in fact, centres of $\bar{3}$ symmetry) lying on the surfaces.

Apart from the dodecagon that forms the boundary of the generating patch for the surface S (Fischer & Koch 1987), the only possible finite frameworks of straight lines capable of being spanned by a portion of a TPMS with cubic symmetry are the configurations of edges of the saddle polyhedra illustrated in Figure 9.23. Observe that the faces of these polyhedra are generating patches for TPMBS: P and C(P) for 9.23a, D and Schoen's C(D) for 9.23b, D for 9.23c, D and P for 9.23d. Other TPMS correspond to spanning the edges of the saddle polyhedra in more complicated ways (Figure 9.9).

Plate XXI (a) Two nonagon generating patches in 1/8 unit cell of the surface C($\pm$Y) and (b) two nonagons spanned by a 'catenoid-like' generating patch of $\pm$Y.

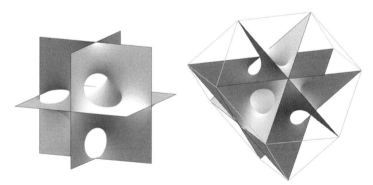

Figure 9.9 Two examples of complicated generating patches obtained by spanning the frameworks of edges of saddle polyhedra in unusual ways.

9.5 Fundamental Patches

As was mentioned in Chapter 3, the *International Tables for Crystallography* (Hahn 1995) specifies an 'asymmetric unit' for each space group G – a polyhedral region that tiles E_3 when repeated by the group. Triply periodic networks or triply periodic surfaces, for example, can be studied and classified in terms of the portion of the structure that lies within an asymmetric unit. This method seems to have been not much used in the case of networks (but see Schoen 1967; Goetzke & Klein 1991). Triply periodic surfaces can be specified by describing the *fundamental patch* within an asymmetric unit. In many cases the faces of an asymmetric unit are either reflection planes or faces partitioned by a diad axis. In such cases a fundamental patch for a TPMS is a polygon bounded by the diads and geodesics lying in the mirror planes (Fogden 1994; Fogden & Haeberlein 1994; Lord 1997). In Figures 9.3–9.6 the fundamental patches are the triangles into which the generating patches have been partitioned.

The software 'Surface Evolver', developed by Ken Brakke and offered free for downloading from his web site, is a very versatile tool for investigating minimal surface configurations (Brakke 1992; 1996; Mackay 1994). In particular, it can find (using a finite element method) the minimal patch bounded by straight lines and mirrors and can employ transforms to display larger portions of the resulting surface. It has been extensively employed for the study of TPMS (Karcher & Polthier 1996; Lord 1997; Mackay 2000; Lord & Mackay 2003).

Fogden & Hyde (1992a, b; 1993) have generalised Schwarz's method and analysed many TPMS, and new varieties have been discovered by their methods (Fogden 1994; Fogden & Haeberlein 1994). The starting point is a knowledge of the normals and angles at the vertices of a fundamental patch bounded by straight edges and reflection planes. We outline briefly their method. Observe that a surface in E_3 can be mapped to a sphere S_2 by mapping each point p of the surface to the point on the unit sphere with position vector $\mathbf{n}$, the unit normal to the surface at p. This is the *Gauss map*. The sphere is then stereographically mapped on to a plane, which is then regarded as the complex plane coordinatised by $\omega = u + iv$. This procedure, for the case of the hexagonal patch of D, G or P (Figure 9.7) is illustrated in Figure 9.10. Observe that a circuit around the perimeter of the hexagonal patch on the minimal surface is represented on the sphere as two circuits around the perimeter of the triangular quadrant and that, equivalently, the angle $\pi/6$ at the vertex C of the triangular fundamental patch becomes $\pi/3$ at the point C on the sphere. This corresponds to the fact that the point C on the minimal surface is a *flat point* (Koch & Fischer 1990) and the point C on in Figure 9.10b is a *branch point* of the Weierstrass function. (A flat point on a surface is a point where the Gaussian curvature is zero.) The orders of the flat points on a minimal surface and the orders of the branch points of its Weierstrass function correspond. Finding the Weierstrass

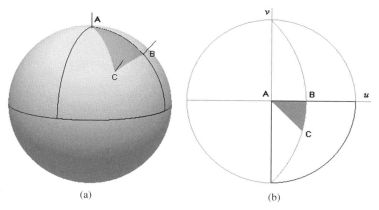

(a)   (b)

Figure 9.10 (a) The Gauss map of the surface patch. The resulting image is the same for P, G and D, because the surface normals are the same in all three cases. (b) A stereographic projection of the Gauss map image. The Weierstrass function for D (or G or P) has a branch point at C.

function of a TPMS is essentially a matter of identifying and classifying its flat points and the normals at those points, and then constructing the analytic function with branch points of the specified orders, at the specified positions.

9.6 Orthorhombic, Rhombohedral and Tetragonal Variants

The Bonnet transformation produces a one-parameter family of minimal surfaces from a given minimal surface. With rare exceptions (such as the trio D, G and P) the resulting surfaces have highly complicated self-intersections. Nevertheless, the boundaries of the fundamental patches of a TPMS and its adjoint ($\theta = \pi/2$) are related in an interesting way: edge lengths are maintained, and straight edges are changed to edges lying on mirror planes and vice versa. Karcher has exploited this property of adjoint fundamental patches and developed a method (the 'conjugate contour method') of obtaining useful information about the metrical properties of a minimal surface, even when the adjoint surface is self-intersecting (Karcher 1989a, b; Karcher and Polthier 1996).

A different kind of family is obtained by changing the proportions of the unit cell. The TPMS obtained in this way are *topologically* equivalent but have quite different metrical properties. Examples of orthorhombic variants of the P surface are shown in Figure 9.11.

Every TPMS with tetragonal, trigonal or hexagonal symmetry exists in an infinite number of variants depending on the *c/a* ratio of the underlying lattice – always with a critical value of *c/a* beyond which no minimal surface exists since two boundaries cannot be spanned by a 'catenoid-like' surface if they are too far apart (Lidin 1988). For a *c/a* ratio below the critical value there are two metrically distinct

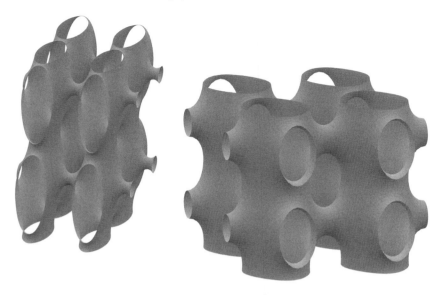

Figure 9.11 Orthorhombic variants of the P surface.

TPMS with the same space group symmetry and topology; the catenoidal-like tunnels with the narrower neck are less stable.

In particular, every TPMS with *cubic* symmetry is a member of two infinite families; a family of tetragonal surfaces obtained by rescaling the lattice along a [100] direction and a family of rhombohedral surfaces obtained by rescaling along a [111] direction. For example, observation of the two catenoid-like patches spanning a pair of equilateral triangles, Figures 9.3e and 9.5d, demonstrates that D and P are two special members of a family of TPMS, which in general have rhombohedral symmetry. This is the family rPD, with symmetry $R\bar{3}m$-$R\bar{3}m$ (2c). The c/a ratio for the special cases D and P are respectively $1/\sqrt{6}$ and $(\sqrt{2})/4$.

A fascinating phenomenon occurs for the family rG (or rY*) of rhombohedral TPMS to which the gyroid belongs (Fogden *et al.* 1993). Figure 9.12 shows the content of a unit cell (in the hexagonal coordinate system) for a few of these surfaces. As the c/a of this cell is increased the curves of intersection of the surface with the planes $z = 0$ and $z = 1/2$ become straight lines. The surface at this stage has become a surface generated by catenoids spanning planar nets of equilateral triangles. In other words, the families rPD and rG have a common member. Thus, the surfaces D, G and P are interconvertible through continuous deformations that have nothing to do with the Bonnet transformation. All intermediate surfaces in the deformations are TPMS without self-intersection. This observation is presumably relevant for a proper understanding of the phase transitions known to take place in lipid bilayers (Hyde *et al.* 1984).

A further interesting phenomenon (Lidin & Larsson 1990) has been shown to occur when the Bonnet transformation is applied to a Schwarz surface H: for a

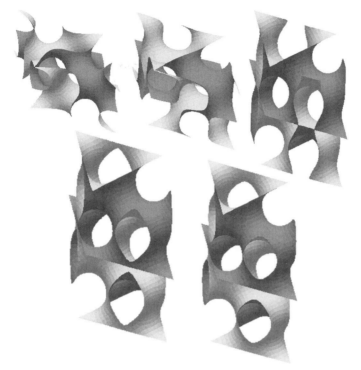

Figure 9.12 A sequence of TPMS leading from the gyroid to a rhombohedral variant of D (or P). Illustrated are the portions of the surfaces within a unit cell of the underlying hexagonal lattice.

particular *c*/*a* ratio the family of Bonnet transformations contains a member of the family rG.

A fuller description of the interrelations among the well-known TPMS, and tetragonal and rhombohedral families of TPMS, has been given by Fogden & Hyde (1999).

9.7 Minimal Surfaces and the Hyperbolic Plane

TPMS can be mapped conformally on to the hyperbolic plane H_2. Figure 9.13 represents the tiling of H_2 by the fundamental regions of the hyperbolic group *642, of H_2, in the Poincaré representation (Coxeter 1989). The Poincaré representation of H_2 is conformal (i.e. angles are correctly represented). The regular tilings 6^4 and 4^6 are obtainable from this figure in an obvious way.

H_2 can be mapped conformally to the minimal surfaces D, G and P so that the fundamental triangles in H_2 are mapped to the fundamental patches (Sadoc & Charvolin 1989). This mapping has been employed in the study of TPMS by various authors (Hyde 1991; Oguey & Sadoc 1993; Hyde & Ramsden 2000a, b; Hyde & Ramsden 2003).

The mapping from the hyperbolic plane to a TPMS is many-to-one; certain finite symmetries of the minimal surface (an obvious example being the fourfold rotations of the P surface – see Figure 9.4) are represented in H_2 by hyperbolic translations. The symmetry groups of D, G and P (respectively $Pn\bar{3}m$, $Im\bar{3}m$ and $Ia\bar{3}d$) are quotient groups of the hyperbolic symmetry group of the tiling pattern of Figure 9.13a.

The many-to-one nature of the mappings from H_2 to a TPMS have been exploited by Hyde & Ramsden (2000b) in the discovery of interwoven three-dimensional (3D) nets. Figure 9.13b and c indicate two ways of inscribing an infinite number of congruent 3-connected trees (with $2\pi/3$ coordination angles) on H_2. In neither of these patterns do the trees intersect each other. Every hexagonal tile centre is a tree node. When the system of trees indicated in Figure 9.13b is mapped to D we get *four* interwoven (10, 3)-a nets, and when the system indicated in Figure 9.13c is mapped to P we get *eight* interwoven (10, 3)-a nets. These are the same configurations that we saw inscribed on the surfaces of the Coxeter–Petrie polyhedra in Figure 7.24. (The polyhedron {6, 6} has the same G–H symmetry and the same topology as D and the polyhedra {6, 4} and {4, 6} have the same G–H symmetry and the same topology as P.)

The mapping of the trees in Figure 9.13b on the gyroid is rather curious. We get two interwoven 3D nets with *curvilinear* edges (Figure 9.14). All angles at the nodes are $2\pi/3$ as in Wells' (10,3)-a. However, if the edges are straightened, keeping the vertices fixed, the angle becomes the tetrahedral coordination angle 109.4...°. (The configuration is obtainable from the double-diamond configuration (Figure 8.22) by systematic elimination of one in four of the edges.) The two interwoven (10, 3)-a nets are directly congruent (either both right-handed or both left-handed); this is in contrast to the pair of interwoven labyrinth graphs of the gyroid, which are enantiomorphic.

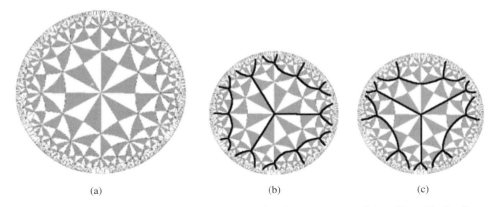

(a)   (b)   (c)

Figure 9.13 (a) The fundamental regions for the symmetry of the tiling (6, 4) of the hyperbolic plane. In the orbifold notation the symmetry group is *642. (b) and (c) Two configurations of 3-connected trees on H_2.

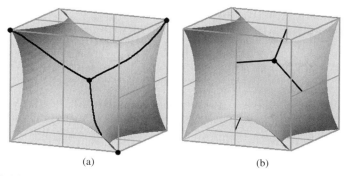

(a)   (b)

Figure 9.14 (a) A portion of the curvilinear (10, 3)-a net on the gyroid. Shown is an eighth of a unit cell of $I4_132$; the whole configuration – a pair of interwoven nets – is generated by the diad rotations about the axes marked on the cube faces. (b) A portion of the gyroidal labyrinth graph.

The same technique also reveals that a trio of interwoven (8, 3)-c nets (Figure 7.23) can be inscribed on an H-surface (Hyde & Ramsden 2000b).

9.8 TPMS with Self-Intersections

Four of the six TPMBS analysed by Stessmann (1934) have lines of self-intersection. Stessmann, however, did not draw attention to this interesting property; the first detailed discussion of a self-intersecting TPMBS was given by Schoen (1970), who discovered a curious *non-orientable* (i.e. one-sided) surface (Figure 9.15). Recently, Elke Koch and Werner Fischer have investigated TPMBS with rectilinear

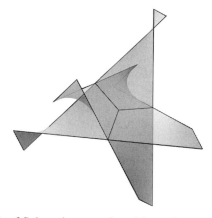

Figure 9.15 A patch of Schoen's non-orientable surface, compose of six pentagonal generating patches. The straight lines are all diad axes belonging to the symmetry group $I4_132$.

self-intersections in a thoroughly systematic way and have found over 70 such objects (Fischer & Koch 1996b; Koch & Fischer 1999; Koch 2000).

A straight line embedded in a minimal surface is *ipso facto* an axis of 180° rotational symmetry. In the neighbourhood of a straight line of self-intersection, therefore, two branches of the surface may pass through each other, cutting each other at an angle $\pi/4$ in which case the line of self-intersection is an axis of four-fold symmetry. Or three branches of the surface may cut each other at an angle $\pi/6$ in which case the line of self-intersection is a sixfold rotational symmetry axis. A line of self-intersection may lie on mirror planes of the symmetry group in which case the number of branches of the surface that pass through it is doubled.

A further possibility is for a minimal surface to have a *branch point* at the intersection of three lines of self-intersection meeting at an angle $2\pi/3$. Schoen's non-orientable surface is of this kind. It has a pentagonal generating patch one of whose

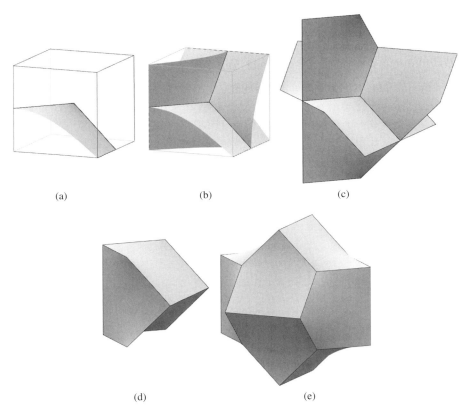

(a)   (b)   (c)

(d)   (e)

Figure 9.16 The surface WI-00. (a) A fundamental patch in an eighth of a unit cell the faces of the cube are mirrors. (b) Six fundamental patches produce a branch point. (c) Six pentagonal generating patches. The edge opposite the $2\pi/3$ angle of the pentagon is also a line of self-intersection. Tiling by the two polyhedra (d) and (e) produces the surface.

angles is $2\pi/3$. Observe that the lines of self-intersection terminate at the branch point (Figure 9.15). The lines of self-intersection constitute a (10, 3)-a net.

A TPMS without self-intersections partitions E_3 into two labyrinthine regions. When lines of self-intersections are present a much richer variety of possibilities arise. The surface may partition space into two, four or even eight 3-periodic labyrinths, it may cut the space into singly periodic tubes, doubly periodic layers or finite cells, or various combinations of these possibilities. We illustrate by just two examples (Fischer & Koch 1996b).

The surface WI-00 (Figure 9.16) has a pentagonal generating patch and produces a tiling of E_3 by two saddle polyhedra with pentagonal faces. The centres of the large tiles (which are distorted pentagonal dodecahedra) lie at the vertices of a lattice complex I and the centres of the small tiles lie at the vertices of a lattice complex W. Fischer and Koch observed that these are respectively, the positions of the Si and Cr atoms in the $CrSi_3$ structure.

The surface WI-10 (Figure 9.17) is closely related to the WI-00 surface. Its generating patch is an octagon, and the space is partitioned into finite cells and tubes. The

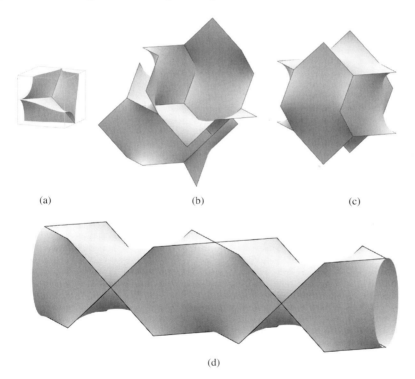

(a)   (b)   (c)

(d)

Figure 9.17 The surface WI-10. (a) Six fundamental patches in a mirror box. (b) Six (octagonal) generating patches surrounding a branch point. (c) The finite cell of the partitioning of space by WI-10 and (d) the infinite tubular region. (The axes of the tubes run along three mutually orthogonal families of parallel diad axes of the symmetry group $Pm\bar{3}mn$.)

finite cell is actually identical to one of the two saddle polyhedra that give the 3D tiling illustrated in Figure 9.25, in which two kinds of octagonal faces occur. Observe that if all the octagons of one type are removed from this tiling the remaining structure is a P surface; removing the octagons of the other kind leaves a WI-10 surface!

WI-00 and WI-10 both have symmetry Pm$\bar{3}$n. WI-10 is orientable but WI-00 is not.

Elser (1996) has described an interesting configuration that minimally spans the set of threefold axes of the space group Ia$\bar{3}$d. These axes do not intersect each other. (They are the axes of the cylinders in the rod packing shown in Figure 4.15b.) Instead of allowing the surface to continue through a threefold axis, the axis is a boundary where three surfaces meet at an angle $2\pi/3$ (as in a minimal foam). The configuration partitions E$_3$ into *three* triply periodic labyrinths. The surface twists around the threefold axes, as in the traditional Archimedean screw, a

Figure 9.18 A present day triple Archimedean helix, such as Archimedes himself could have used, in the museum of a salt farm in Sicily.

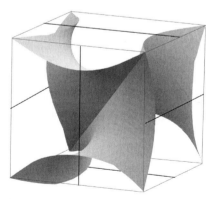

Figure 9.19 The portion of Elser's 'Archimedean screw' in an eighth of a unit cell of Ia$\bar{3}$d.

Plate XXII A stereo view of a unit cell of Elser's Archimedean screw.

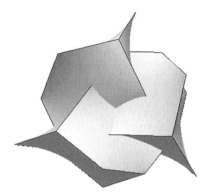

Figure 9.20 Three interpenetrating saddle polyhedra associated with three interwoven (distorted) D nets.

device for pumping water (Figure 9.18). Figure 9.19 shows the portion of Elser's 'Archimedean screw' configuration in an eighth of a unit cell. The whole configuration is generated by applying the twofold rotations whose axes are marked on the cube faces. Plate XXII is a stereo view of a unit cell, viewed in a direction close to the direction of a threefold axis.

The labyrinth graph for each of the three labyrinths turns out to be a D net tetragonally deformed (by a factor $1/\sqrt{2}$ along a $\bar{4}$ axis). The three interwoven D nets are indicated in Figure 9.20, in which just one saddle polyhedron (four skew hexagons) from each net is shown.

9.9 Tiling of Triply Periodic Surfaces

Quite a few instances of tiling patterns on TPMS have already been encountered, in this and previous chapters. In particular, the generating patch for a triply periodic surface generates the surface along with a monohedral tiling of the surface. The (10, 3)-a nets on D, G and P are particularly striking examples of structures in 3D-Euclidean space that can be interpreted as configurations in on D, G and P are particularly striking examples of structures in 3D-Euclidean space that can be interpreted as configurations in 2D hyperbolic space.

In Chapter 2 we presented some topological identities for those tilings on a surface of genus g for which every vertex has the same connectivity. The application of these formulae to triply periodic surfaces requires a definition of *genus* for such a surface. The portion of the surface in a unit cell is an obvious choice of definition. The genus can be calculated from any arbitrary tiling of this portion, according to $g = 1 - \chi/2, \chi = V - E + F$. It is expedient to use the *primitive* unit cell of the underlying lattice, and for a *balance* surface, the primitive unit cell for the subgroup H (the labyrinth symmetry). This is the choice adopted by Fischer and Koch. (For cubic symmetry groups, Gozdz & Holyst (1996) adopt a different convention, defining the genus from a cubic unit cell.) The genus of a triply periodic surface can also be calculated from the topological characteristics of its labyrinth graphs (Fischer & Koch 1989b). A fractional Euler characteristic can be obtained from a single fundamental patch, by taking $V = \theta/2\pi$, where θ is the sum of the angles at the vertices of the patch, E is half the number of edges (because each edge is shared by two patches) and $F = 1$ (Lord 1997).

The genus is 3, for D, G and P.

For a 3-connected tiling of any of these surfaces, we have

$$3F_3 + 2F_4 + F_5 - F_7 - 2F_8 \cdots = 6(2 - 2g) = -24.$$

Thus, a tiling of these surfaces by a 3-connected net of hexagons (e.g. a graphite sheet) is possible if *12 octagonal* rings, per primitive unit cell, are included among the hexagons (Mackay & Terrones 1991). Figure 2.21 showed such a tiling on two unit cells of the P surface. As for the fullerenes, a wide variety of hypothetical 'Schwarzites' can be deduced by application of the iterative procedures described in Section 2.11. More intricate 3D nets derived from tilings of P, D and IWP have been derived and illustrated by Mackay & Terrones (1993). A model for amorphous carbon based on curved graphite sheets has been proposed (Townsend *et al.* 1992).

Wells (1977) proposed and illustrated several polyhedra {7, 3}. Three varieties exist as tilings of P, G and D by heptagons – with 24 heptagons per (primitive) unit cell. They can be obtained by mapping from H_2, as indicated in Figure 9.21. Note that this heptagonal tiling of H_2 is *not* the regular 7^3 tiling of H_2 (Figure 9.22). The tiles are congruent, all angles are $2\pi/3$, but the (hyperbolic) edge lengths are not all equal. The dual polyhedron {3, 7} on D is related to an interesting polyhedral net

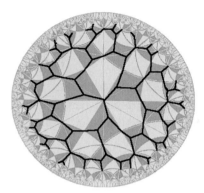

Figure 9.21 A 7^3 tiling of the hyperbolic plane superimposed on a regular 4^6 tiling. By mapping the quadrilaterals to pairs of fundamental patches of D, G or P will give a heptagonal tilings of these surfaces.

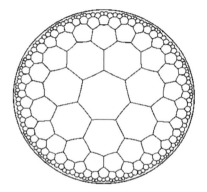

Figure 9.22 The regular tiling 7^3 of the hyperbolic plane.

constructed from icosahedral nodes linked by octahedral tunnels. Its surface has seven *equilateral* triangles around every vertex (Plate XV). The relation between zeolite frameworks and tilings of H_2 was recognised by Blum *et al.* (1988). Ingenious descriptions of a number of naturally occurring network structures – zeolites, silicates and clathrate hydrates – in terms of tilings of TPMS have been deduced by Steven Hyde and Sten Andersson. The surface tilings are in turn related to tilings of the hyperbolic plane (Andersson & Hyde 1984; Hyde & Andersson 1985; Hyde 1991; Hyde & Ramsden 2000a, b).

9.10 Saddle Polyhedra

Polyhedra with straight edges but with faces that are not planar are *saddle polyhedra* (Schoen 1968). A non-planar face can be taken to be a minimal surface patch spanning a skew polygon formed by a circuit of edges. Williams (1979) has presented several interesting ways of obtaining saddle polyhedra from the Platonic and Archimedean solids, for example, by radially translating the mid-points of its edges, a p-gon face becomes a skew 2p-gon. Pearce (1978) has developed a constructional system (the Universal Node system) consisting of nodal connectors with cubic symmetry and rods with lengths in the ratios $1 : \sqrt{2}: \sqrt{3}$, attached respectively along fourfold, twofold and threefold axes, and illustrated how a large number of saddle polyhedra can be produced. Especially interesting are the saddle polyhedra that can be fitted together to produce a tiling of E_3. We illustrate in Figures 9.22–9.25 some particularly fascinating examples of 3D tilings by saddle polyhedra, involving either a pair of tiles, or a single tile.

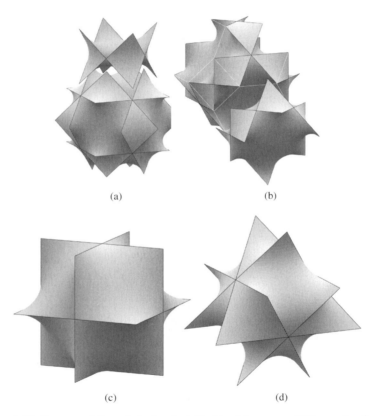

(a)   (b)

(c)   (d)

Figure 9.23 Some saddle polyhedra that tile E_3: (a) a space filling pair of saddle polyhedra whose faces are quadrilateral patches of P and octagonal patches of C(P), (b) A space filling pair whose faces are quadrilateral patches of D and dodecagonal patches of C(D) that produce a space filling, (c) a space filling polyhedron whose faces are hexagonal patches of D and (d) hexagonal patches of P and quadrilateral patches of D. This case is particularly remarkable in that this single tile can fill a labyrinthine half-space in two distinct ways, with totally different symmetries – namely, the labyrinths of P and the labyrinth of D.

The framework of edges of the largest of the two polyhedra illustrated in Figure 9.23a consists of the diagonals of the hexagonal faces of a truncated octahedron. The faces are skew quadrilaterals and skew octagons. The figure suggests the way the pair of polyhedra can be fitted together to produce a space filling. The quadrilateral faces join up to produce a P surface, and the octagonal faces similarly build up a C(P) surface. The pair of saddle polyhedra in Figure 9.23b also tile E_3. The edges of the large one are the diagonals of all the square faces and four of the

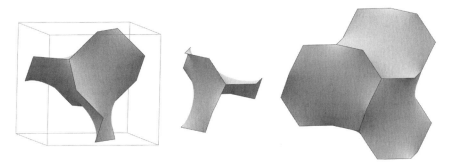

Figure 9.24 Two views of a curious space filling trihedron with three skew decagon faces, and a cluster of three of them. In the resulting tiling of E_3, the edges form a gyroidal labyrinth graph (Wells' (10, 3)-a net).

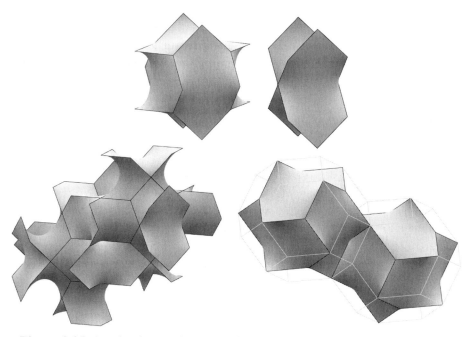

Figure 9.25 Another interesting space filling pair. Faces are two kinds of skew octagons. The net formed by the vertices and edges is the Pt_3O_4 net.

hexagonal faces of a truncated octahedron. The two continuous surfaces in this case are Schwarz's D surface and Schoen's C(D) surface. The hexagonal faces of the polyhedron in Figure 9.23c are *Petrie polygons* of cubes (a Petrie polygon is a circuit of edges so that every 2, but no 3, consecutive edges belong to a face). The faces are patches of the D surface. The hexagonal faces are Petrie polygons of cubes. The faces of the tile shown in Figure 9.23d are quadrilateral patches of the D surface (Petrie polygons of tetrahedra) and hexagonal patches of P (Petrie polygons of octahedra). Its edges are the diagonals of the hexagonal faces of a truncated tetrahedron. These four tilings of E_3 have all been described and illustrated by Pearce (1978). They are particularly important in the theory of TPMS (Lord & Mackay 2003), because the embedded straight lines in the minimal surfaces P, D, C(P) and C(D) are the edges of the tiles in the space fillings. Particularly fascinating is that the labyrinths of these surfaces are produced by tiling only half of space. Figures 9.24 and 9.25 serve to indicate the variety of possibilities for tiling space with saddle polyhedra.

Stephen Hyde has described intriguing relationships between 3D tilings by saddle polyhedra (and the related minimal surfaces) and the atomic configurations of a variety of material structures (Andersson & Hyde 1984; Hyde *et al.* 1984; 1996).

9.11 Other Kinds of Triply Periodic Surfaces

A TPMS can be characterised by space group symmetry and topology. One may consider non-minimal surfaces with the same symmetry and topological characteristics. Obvious instances are triply periodic polygonal surfaces of various kinds. Considering only *smooth* surfaces, an obvious next step after considering TPMS – equivalently, surfaces of zero mean curvature – is to consider surfaces of *constant* mean curvature. A 'soap film' spanning a bounding curve, as in 'Plateau's problem' takes the form of a surface of constant mean curvature if there is a difference in pressure between the two sides of the film.

Theoretically, every minimal surface S belongs to a one-parameter family of surfaces of constant mean curvature, which are the surfaces normal to a field of unit vectors satisfying div **n** = 0 and coinciding with the normal to S, on S. For practical purposes, however, this deceptively simple prescription is quite intractable. However, Surface Evolver, which has proved invaluable for the investigation of minimal surfaces, can deal readily with triply periodic surfaces of constant mean curvature, by imposing a volume constraint on a labyrinth. Figure 9.26 shows a 'double-diamond' surface. Its two branches are separated by a minimal D surface. It can be visualised as two systems of tubes enclosing the labyrinth graphs.

Further important examples of triply periodic surfaces are the Fermi surfaces of metals (Harrison 1961; Mackintosh 1963), and the *equipotential surfaces* associated with a triply periodic distribution of positive and negative charges as in an ionic crystalline material. A simple example is the CsCl strucure – two kinds of ions, at

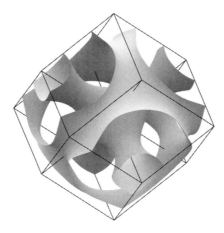

Figure 9.26 A portion of a double-diamond surface – a surface of constant mean curvature. The faces of the bounding rhombic dodecahedron are reflection planes.

(000) and (1/2 1/2 1/2). The zero potential surface closely resembles the minimal surface P. Similarly, the NaCl structure has an equipotential surface that looks like the Neovius surface C(P).

A technique for generating triply periodic surfaces has been developed, and applied to the theoretical simulation of the separation surfaces that occur in block copolymers, in which two regions occupied by two materials form labyrinthine configurations. The method employs the *structure factors* that have long been known and used by crystallographers and are fundamental in the Fourier transform techniques of X-ray crystallography. The surfaces obtained in this way, the *nodal* surfaces and *level* surfaces, will be the subject of the next section.

9.12 Nodal Surfaces and Level Surfaces

Let x, y and z be coordinates referred to the underlying lattice of a crystal. A function

$$F_{hkl} = \cos(hX + kY + lZ - \alpha)$$

($X = 2\pi x$, $Y = 2\pi y$, $Z = 2\pi z$) describes a plane wave in the direction (hkl). These functions are summed over all identical atoms in the unit cell at positions (x, y, z) that are equivalent for the space group. Finally, a weighted summation over inequivalent atoms in the crystal gives the amplitudes and phases for the diffraction pattern produced by reflection from the (hkl) planes.

To generate triply periodic surfaces, the functions F_{hkl} are summed over all symmetry equivalent values of h, k and l to obtain a function $f_{hkl}(x, y, z)$ that defines a *nodal surface* $f_{hkl} = 0$ and its associated *level surfaces* f_{hkl} = constant (von Schnering & Nesper 1991; Schwarz & Gompper 1999).

For small values of h, k and l nodal surfaces can be obtained that are remarkably close to known minimal surfaces with the same space group symmetry. An indication of the accuracy of the D, G and P surfaces, regarded as approximations to the analogous minimal surfaces has been assessed by Lambert *et al.* (1996) in terms of area per unit volume. Larger values of f_{hkl} produce more intricate surfaces with high genera. Von Schnering & Nesper (1991) have provided a list of the functions that correspond to some of the well-known minimal surfaces. Mathematica is useful for displaying them. The table below gives a few examples from von Schnering and Nesper. The curious nodal surface shown in Plate XXIII has the same G–H symmetry as the gyroid, but a slightly more intricate topology.

Analogous TPMS	Space group H	hkl	$f(x, y, z)$
D	$Fd\bar{3}m$	111	$\cos X \cos Y \cos Z + \sin X \sin Y \sin Z$
G ($=Y^*$)	$I4_1 32$	110	$\sin Y \cos Z + \sin Z \cos X + \sin X \cos Y$
P	$Pm\bar{3}m$	100	$\cos X + \cos Y + \cos Z$
C(P)	$Pm\bar{3}m$	100	$\cos X + \cos Y + \cos Z$
		111	$+ 4\cos X \cos Y \cos Z$
IWP	$Im\bar{3}m$	110	$2(\cos Y \cos Z + \cos Z \cos X + \cos X \cos Y)$
		200	$-(\cos 2X + \cos 2Y + \cos 2Z)$

Every nodal surface belongs to a family of level surfaces (surfaces of constant f). For nodal surfaces that approximate minimal surfaces, the associated level surfaces are approximations to the associated surfaces of constant mean curvature. Plate XXIV is a level surface associated with the gyroidal nodal surface Y**.

Wohlgemuth *et al.* (2001) have studied the effect of combining the functions in linear combinations to obtain sequences of nodal surfaces and level surfaces. A nodal surface analogous to Schoen's OCTO for example belongs to the same one-parameter family as a nodal P and a nodal IWP. This possibility of transforming a nodal surface into a different nodal surface by continuous variation of a parameter has been applied to the elucidation of transitions in network-based mineral structures, by Leoni & Nesper (2000).

Gozdz & Holyst (1996a, b, c) have obtained triply periodic surfaces of high genus as the local minima of the Landau–Ginzburg Hamiltonian for microemulsions. In particular, they obtained a family of higher genus surfaces with Ia3d symmetry, the first member of which is the Schoen's gyroid. (Beware that the definition of genus adopted by Gozdz and Holyst is based on the cubic unit cell rather than the primitive cell.) Schwarz & Gompper (1999) employ the Fourier series approach to approximate nodal surfaces and compute Gaussian and mean curvatures.

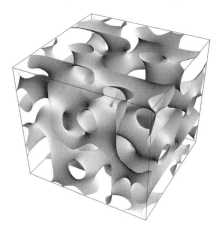

Plate XXIII von Schnering and Nesper's C(Y**), formula $3(\sin X \cos Y + \sin Y \cos Z + \sin Z \cos X) + 2(\sin 3X \cos Y + \sin 3Y \cos Z + \sin 3Z \cos X - \sin X \cos 3Y - \sin Y \cos 3Z - \sin Z \cos 3X) = 0$. This nodal surface has the same symmetries as the gyroid, I43d–I4$_1$32. Whether a minimal surface exists with this topology and these symmetries is not known.

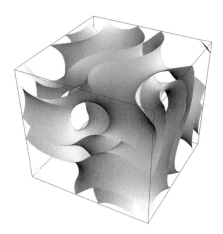

Plate XXIV A 'double gyroid'. A nodal surface with the symmetry (I4$_1$32) and topology of the family of constant mean surfaces to which the gyroid belongs. The formula is 0.8 $(\sin 2X \sin Y \cos Z + \sin 2Y \sin Z \cos X + \sin 2Z \sin X \cos Y) - 0.2$ $(\cos 2X \cos 2Y + \cos 2Y \cos 2Z + \cos 2Z \cos 2X) + 0.33 = 0$.

Some interesting relationships between rod packings and nodal periodic surfaces have been explored by von Schnering *et al.* (1991). Similarly, triply periodic surfaces and sphere packings are also interrelated (Elser 1994).

Nesper & Leoni (2001) have produced quite beautiful tilings and patterns on nodal surfaces from the variations in the values of higher order structure factors, over a nodal surface given by one of lower order.

10

Novel Atomic Configurations in Metallics

In considering the manifold variety of the geometrical arrangements of atoms, the structures of metals take an important place, because of their physical properties and because they are somewhat alien to the structures of life. We must consider first the nature of metal atoms and the forces which bond them. Metals are characterised by the metallic bond based on free electrons. This leads to a few close-packed structures. The rich variety of atomic configurations seen in the larger class of inorganic materials with covalent and ionic bonding has long remained the envy of metallurgists. However, over the past few decades non-equilibrium processing has led to new configurations of quasicrystalline and non-crystalline phases, rivalling in complexity other inorganic structures. These studies have also revealed deep connections between the structure of intermetallics and clathrates and even biological materials with helices.

10.1 Geometrical Considerations

The ancient Greeks were particularly fascinated by geometry. They sought the most perfect solids and identified five of them (Figure 2.15) which became known as the Platonic solids. They include the tetrahedron, the cube and the octahedron, seen often in the external forms of crystals. They also include the icosahedron and the dodecahedron. These possess the maximum point symmetry with 6 fivefold, 10 threefold and 30 twofold axes of rotational symmetry.

The earliest study of close packing of spheres can be traced back to the Indian astronomer, Aryabhata. In his *Aryabhatiya* he derived formulae for the number of balls in a triangular pile (Shukla 1976). The resurgence of interest in this subject must be ascribed to another astronomer, Johannes Kepler. The Platonic solids fascinated Kepler, who saw in them harmony on a planetary scale. From cosmology he moved to the microcosm and in 1611 speculated on the six-pointed snowflake and attributed the symmetry to the underlying atomic configuration (Kepler 1611).

This concept has been hailed as 'the first recorded step towards the mathematical theory of the genesis of inorganic or organic form'. The elucidation of the periodic arrangement of atoms became a major preoccupation of crystallographers. It was possible to demonstrate that there are exactly 14 Bravais lattices, 32 point groups and 230 space groups.

Kepler also discussed the close packing of spheres and stated without proof that the cubic close packing (ccp) leads to maximum efficiency in space filling (Kepler 1611). His conjecture became famous; as observed by Rogers (1958), 'Many mathematicians believe and all physicists know' that this is so. In 1900, Hilbert made this conjecture part of his 18th problem. This conjecture lasted nearly five centuries till Hales found a proof in 1997.

10.2 Pure Metals

Most elements are metallic. Almost all the remaining elements can be made metallic (Rao & Gopalakrishnan 1997; Cottrell 1998). The metallic bond is characterised by free electrons and is non-directional. The atoms in pure metals tend, therefore, to be close packed. Two common variants of the densest close packing (with coordination number (CN) 12 and packing fraction $0.74048 = \pi\sqrt{18}$) can be distinguished, ccp, with sphere centres on a face-centred cubic (fcc) lattice and hexagonal close packing (hcp). The coordination polyhedra for these two structures are the cuboctahedron and the diheptahedron (actually a twinned cuboctahedron – illustrated in Figure 10.1). Another simple arrangement found in many pure metals is the body-centred cubic lattice (bcc), which has CN 14 and a slightly lower atomic packing density $0.68017 = (\pi\sqrt{3})/8$. Steurer (1996) has surveyed the structure of metallic elements; 58 metals possess a close-packed arrangement (either cubic or hexagonal) and 23 metals crystallise in the bcc structure.

Most metallurgists are conversant with these three simple structures. A few metals have other crystal structures and many metals can show other structures, or can be polymorphic, with the application of high pressures. They may also produce more complex structures on very rapid cooling from the vapour or melt.

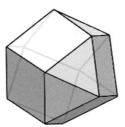

Figure 10.1 The disheptahedron.

Both diffraction (X-ray, electron and neutron) and imaging techniques (field-ion microscopy, high-resolution transmission electron microscopy (TEM) and high-angle annular dark field, HAADF) have been used to study the atomic configuration in metals and alloys. Figure 10.2 shows a field-ion micrograph of platinum, where individual atoms can be resolved (Hren & Ranganathan 1968; Mueller & Tsong 1969). It also gives the information that platinum crystallises on an fcc lattice. It is interesting to observe that the prominence of planes follows the order (111), (200), (220), etc. This is the same sequence observed in a Debye–Scherrer X-ray pattern and demonstrates that in both cases the interplanar distance is the key factor, as first pointed out by Moore & Ranganathan (1967).

A few metals crystallise in other structures. Many show polymorphism and many more can display polymorphism with a variation in pressure. This structural variation can often be correlated with the electronic configuration. In transition metals and noble metals the tight binding approximation may have to be used, as s–d electron hybridisation occurs and results in directional bonding. An interesting example of the complexity which may occur in metallic structures is provided by α-manganese (which was described in Section 6.5). This has a cubic structure with 58 atoms per unit cell and it appears that atoms of the same element display three different sizes, so that this structure must be regarded as a case of self-alloying by a metal with the structural characteristics of a ternary intermetallic compound.

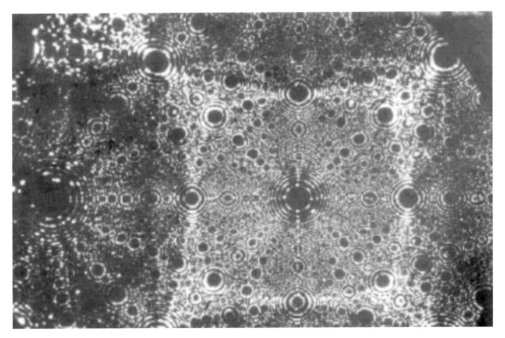

Figure 10.2 Field-ion micrograph of platinum. Note the order of prominence of (111), (200) and (220) planes.

In general, metal atoms are characterised by their radii and by their electronegativity, that is, the ease with which they donate free electrons to the collective electron cloud of a block of metal. Tables of these properties are readily available. It is these free electrons which give metals their most characteristic property, that of conducting electricity, but the many sophisticated properties of metals cannot be adequately dealt with without using quantum mechanical considerations. Electrons, and indeed all matter, are both waves and particles. However, in many cases the geometrical structure of metals is mostly determined by the closest packing of positively charged spherical atoms in a cloud of negative electrons. In recent years the ultra-rapid cooling of metallic melts has produced a great variety of much more complex arrangements, the chief among these being those of quasicrystals.

The principal kinds of interatomic bonds are covalent, electrovalent or ionic, metallic and the hydrogen bond. Covalent bonds are strongly directional and depend on the nature of the atoms involved and they have a preferred length. Ionic bonds are essentially the lines of force between positively charged and negatively charged atoms and are not otherwise directional. They represent problems in classical electrostatics. The metallic bond is more subtle, being essential quantum mechanical, as are the other bonds of course, but is basically non-directional and represents a general compressive force leading to the closest packing of atoms, subject to their characteristic radii and to the possibilities of rearrangement which may depend on the previous history. That is, atoms may be trapped in an unfavourable configuration where the energy of thermal vibrations, which depends on the temperature, is insufficient to carry them over the hump to a configuration of lower energy.

The CN of an atom is usually the number of nearest neighbours, but in some cases it may be necessary to distinguish between bonded (covalently connected) neighbours and those which are simply nearest. The CN of a particular element may be characteristic of the atom and its neighbours.

10.3 Alloys

Alloying leads to a far greater variety of structures. Many of the ordered compounds that are formed can, however, be again be considered as derivatives of the simpler structures. However, research over the past 40 years has revealed that new stable and metastable states of metallic phases can be produced. These include structurally complex intermetallics, non-crystalline and quasicrystalline phases. This leads to an enormous enrichment of possible atomic configurations in the metallic state. There are intriguing links among these different configurations. It is also possible to design microstructures, where these can occur in various profitable combinations with superior properties. Several reviews treat the subject of the structure of intermetallics, albeit with varying perspectives (Pauling 1960; Samson 1968;

1969; Kripyakewitsch 1977; Villars & Calvert 1986; Nesper 1991; Ranganathan *et al.* 1997; Cottrell 1998; Watson & Weinert 2001). The recurring theme of geometrical frustration in forming random and close-packed structures has been explored in elegant terms in Nelson & Spaepen (1989), Venkataraman *et al.* (1989) and Sadoc & Mosseri (1999).

10.4 Solid Solutions

A number of metals dissolve in each other forming solid solutions. When a solute atom is much smaller than the solvent atom, it may dissolve interstitially occupying the voids in the parent structure. When the sizes of the two atoms are comparable, substitutional solid solutions are formed. Hume-Rothery (1926) framed empirical rules for the formation of extended solid solutions: the component atoms should not differ more than 15% in diameter; the electronegativity difference must be greater than 0.4; a lower valent metal will dissolve more of the higher valent metal than vice versa; if the crystal structures are the same, extensive solubility may be expected. Cu–Ni, W–Mo, Ti–Zr and Si–Ge form binary isomorphous systems. Substitutional solid solutions can be random or ordered. β-brass in the ordered configuration has the B2 structure – the two kinds of atom alternate on a bcc lattice as in CsCl. These semi-empirical rules have been found to be extremely useful. Alloys that satisfy these criteria are known as *Hume-Rothery phases*. The third rule of lower-valent metal as solvent has many exceptions. When these conditions are not favourable, intermetallics, quasicrystals, structurally complex intermetallics and glasses can be formed. The differentiation of the combination of alloying elements leading to these configurations continues to be subtle and elusive.

Electron concentration plays an important role in determining the structure of Hume-Rothery phases. Alloys with closely related structures but with quite different compositions may be related by having the same ratio of number of valence electrons to number of atoms. The β-phases such as $CuZn$, Cu_3Al and Cu_5Sn for example are all essentially simple bcc structures and all have *electron/atom*, $e/a = 3/2$, the hcp Hume-Rothery alloys are characterised by $e/a = 7/4$, and the γ-phases, which occur in a wide range of compositions but are all describable as a bcc array of 26-atom 'γ-brass clusters' and all have $e/a = 21/13$.

10.5 Intermetallics

Metals can also form intermetallic compounds on alloying. Pauling (1923) was the first to use X-ray diffraction to study the crystal structure of intermetallics. He reported that Mg_2Sn has the fluorite structure. It is instructive to classify the intermetallics into four types: Zintl phases, Hume-Rothery phases, Laves phases and

Hägg phases. They appear as a consequence of the violation of the rules set by Hume-Rothery.

When the constituent atoms have appreciable difference in electronegativity, compounds form obeying normal rules of valence. These are known as *Zintl phases* or valence compounds. Examples are Mg_2Sn, Mg_2Si, Mg_3As_2 and MgSe. In Mg_2Sn, Sn occupies the position of the fcc lattice points while Mg occupies the tetrahedral voids, as in the picture at top left in Plate VI. The Zintl border separates the electropositive and electronegative elements. Compounds formed by taking elements across the border lead to valence compounds following the octet rule. Depending on the difference in electronegativity the intermetallics can thus made to have ionic to covalent bonding. It is also possible to have polyatomic anions and polyatomic cations leading to complex structures. There has been an explosion in research on Zintl phases in recent times.

In a famous and imaginative line of research Hume-Rothery (1926) has shown that some intermetallic compounds – CuZn, Cu_5Zn_8 and $CuZn_3$ – which do not appear to obey normal valence rules can be considered as electron compounds and occur at characteristic *e*/*a* ratios. Corresponding to 3/2, 21/13 and 7/4 we find bcc, γ-brass (complex cubic; section 6.4) and hcp structures.

Bradley & Thewlis (1926) described the γ-brass structure in terms of a cubic unit cell consisting of 27 cubic units – a $3 \times 3 \times 3$ array of a bcc lattice. The sites at the vertices of this block and the one at its centre are vacant, leading to a unit cell with 52 atoms (Figure 10.3). The γ-brass structure was regarded as a distorted version of this geometrical model, in which the atoms are shifted from the exact positions. Bradley & Jones (1933) described it in terms of a cluster of concentric polyhedral shells centred around the vacant sites (Figure 10.4). The first shell is a regular tetrahedron of four atoms. This is surrounded by four more atoms, over its

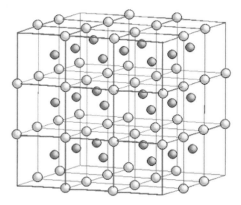

Figure 10.3 Gamma brass, a Hume-Rothery phase, visualised as a $3 \times 3 \times 3$ stack of cubic units.

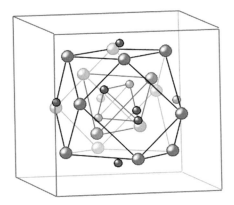

Figure 10.4 The γ-brass cluster as a set of nested polyhedra.

faces, forming a larger tetrahedron. The third shell, of six atoms, is octahedral, one atom over each edge of the small tetrahedron. The fourth is cuboctahedral, with 12 sites. (In the outermost cuboctahedron, the square faces of the Archimedean cuboctahedron are replaced by approximately golden rectangles – the configuration occurs in the polytope {3, 3, 5} with exact golden rectangles as the faces of the outer shell.) The clusters are arrayed in a bcc arrangement. As discussed in Chapter 6, the 26-atom cluster is a polytetrahedral structure that can be seen as a configuration of four interpenetrating icosahedra.

Consideration of size differences plays an important role in the intermediate phases known as Laves phases (see Figure 3.17). These occur with a stoichiometry of AB_2 at a definite radius ratio $r_A/r_B = \sqrt{(3/2)} \sim 1.225$. The A and B elements can be from distant parts of the periodic table or can be neighbouring elements. Examples are $MgCu_2$, $MgZn_2$, $MgNi_2$ and $CaZn_2$. *Hägg phases* – in which the size ratio is extreme – are also known as *interstitial* compounds, where H, B, C, N and O occupy interstitial voids. Hydrogen goes into tetrahedral voids (TiH). Carbon goes into octahedral voids (TiC). Note that TiC has fcc structure, while pure Ti is bcc and hcp.

In deciding on the structure of intermetallics David Pettifor brought new insights. Instead of considering independently the various factors such as size, electronegativity and valence, Pettifor introduced a chemical scale which takes account of the effect of these parameters as well as bond orbital effects. On the basis of this chemical scale Pettifor then assigned a unique *integer* to each element – the *Mendeleev number* (Pettifor 1984). In binary intermetallics this single parameter could be used to differentiate among different structures. With the advent of powerful computers, *ab initio* calculations of structures and phase diagrams are progressing at a furious pace. Nevertheless, the intuitive appeal of Hume-Rothery rules and the phenomenological approach of Pettifor is undeniable.

10.6 Quasicrystals

Shechtman's discovery in 1982 of an alloy of aluminium and manganese exhibiting fivefold rotational axes in reciprocal space (Shechtman *et al.* 1984) challenged many widely held assumptions. This phase, exhibiting icosahedral symmetry, was the first of a host of materials to be discovered having a hitherto unexpected kind of ordered structural arrangement of atoms. They were christened 'quasicrystals' to distinguish them from 'true crystals', which are ordered on an underlying lattice.

All crystals belong to one or other of the 230 crystallographic space groups and are periodic in three independent directions. That is, there is a unit cell (containing one or more units of pattern) which is repeated by translation on a three-dimensional (3D) lattice. The symmetry may involve rotation operations of order 2, 3, 4 and 6 (possibly as inversion or screw axes) but other kinds of rotational symmetry are forbidden. In particular, no *fivefold* axes can occur. (The mathematical proof of this was first given by Leonard Euler.) Thus, diffraction patterns *of crystals* never show the tenfold axes of symmetry (a centre of symmetry is added by the diffraction process).

Textures of many crystallites may be distributed statistically to show any symmetry. In particular, a diffraction pattern from a texture might show axes of tenfold symmetry. Multiply twinned crystals, like snowflakes, may sometimes show tenfold symmetry, and are well known, having been recognised for more than a century. Twinned crystals of native copper with fivefold symmetry are known. More recently (from 1960), with the development of electron microscopy, minute icosahedral particles of gold have been seen and are found to be composed each of 20 crystallites of fcc gold. The 20 tetrahedral gold crystallites need only a slight distortion to fit together to make a regular icosahedron. The minimisation of the surface tension makes such an assembly stable if it is small enough. Diffraction observations, and direct electron microscopy, show the particles to be composed of normal gold crystallites in 20 orientations.

The diffraction patterns of quasicrystals, which showed many discrete spots, as patterns from crystals do, could be distinguished from twinning. The patterns of spots did not correspond to that from a crystal lattice and represented a problem which delayed publication of the findings for two years. The material could thus not be a crystal. What was it? Such material came to be known as 'quasicrystalline' and the icosahedral material, a 'quasicrystal'.

Quasicrystals can have symmetries other than icosahedral. In particular, extensive studies have been done on *decagonal* (tenfold), *octagonal* (eightfold) and *dodecagonal* (twelvefold) quasicrystals. These have one periodic axis and the plane perpendicular to them is quasiperiodic. Thus they are known as '2D quasicrystals' (Ranganathan *et al.* 1997). Among them, decagonal quasicrystals have received maximum attention. Figure 10.5 shows the electron diffraction patterns from a decagonal quasicrystal in Al–Mn alloy (Chattopadhyay *et al.* 1985a).

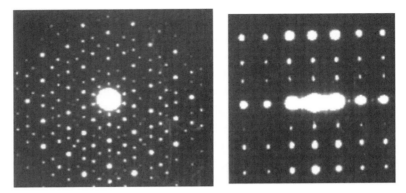

Figure 10.5 Electron diffraction pattern from a decagonal quasicrystal.

Before the announcement in 1984 of the experimental findings, a number of theoretical developments had taken place presaging the discovery. There had been investigations as to how the classical formalism of crystallography might be generalised or widened to include some of the other more general forms of order encountered in the natural world, such as fibres and textures. It had long been recognised, particularly by William Astbury (who coined the term 'molecular biology'), that materials such as wool, cotton and muscle give diffraction patterns, clearly not crystalline but containing a great deal of information. It was an apparent paradox that glass (popularly called 'crystal') gives X-ray diffraction showing only diffuse scattering with no clear spots or bands and little information, while wool gave complex patterns obviously highly informative, if only one had the key. The theory of scattering from a helix developed from the early ideas about the structure of wool and various polymers like polyethylene.

There had already been particular investigations of fivefold symmetry, partly because of the finding of the icosahedrally twinned particles of gold mentioned above, and partly because of the picture of the possible structure of liquids and glasses proposed by J. D. Bernal and F. C. Frank, which suggested that the formation of icosahedral clusters of 13 atoms would prevent the achievement of crystalline order. Local order would here predominate over global order. The observation of virus particles with the symmetry of the icosahedron had also attracted attention. In this case the particles could aggregate into crystals as a compromise between local order and global order.

The relevant investigations included the following:

1. Kramer's ingenious method for dissecting an icosahedron into (3D) tiles of seven kinds, each of which could be itself dissected into smaller versions of the same tiles furnished an important example of *hierarchy* as an alternative to lattice periodicity for solid structures (see Section 5.6).

2. Roger Penrose looked for a set of tiles, with local joining rules (like a jig-saw puzzle), which would tile the plane only aperiodically, that is, preventing the appearance of a lattice. The pattern never repeated itself exactly. He found a set of two tiles, an acute rhombus and an obtuse rhombus which did this (see Figure 2.9). (Other mathematicians, as well as Penrose himself, had already produced patterns for which more than two different tiles were necessary to force aperiodicity.) The pattern can be generated like a jig-saw puzzle, but there is no infallible rule for placing the pieces, it might be necessary to remove many tiles at various stages and proceed again differently. Penrose also gave rules for dissecting the pieces into smaller pieces so that an indefinitely large pattern could be thus generated recursively.

3. Robert Ammann showed that the corresponding dissection of 3D space could be achieved with two tiles which were the acute and obtuse rhombohedra of Kowalewski (1938).

4. Steinhardt *et al.* (1983) examined possible packings of spheres to estimate quantitatively the degree to which icosahedral clusters might occur.

5. Alan Mackay sought hierarchic structures as an alternative to lattice repetition. Packing pentagons leaves gaps – tilings with regular pentagonal tiles with the gaps filled in a systematic way had been studied by Albrecht Dürer and by Kepler. Mackay's version was effectively the Penrose tiling (Mackay 1982). It was shown that this tiling gave a diffraction pattern which later proved to be like that of the alloys. This was part of a programme of Bernal to produce a 'generalised crystallography' which would include a wider range of ordered structures than orthodox crystals. Hierarchic patterns used a redundant set of five axes in the plane or six axes in 3D and corresponding points could be related by integer coordinates with respect to these 'pseudo-lattices'. The redundancy implies that there would be two incommensurable steps in the same direction. (We are already familiar with this phenomenon since the period of the sun (the year) and that of the moon (the month) are incommensurable and cause great difficulties in organising the calendar.)

6. de Bruijn (1981) showed how tilings in two or three or more dimensions could be constructed from 'multigrids', the intersections of which showed the order in which various tiles, determined by the nature of the intersection, should be laid. The Penrose tiling appeared as a special case.

The theoretical machinery was thus ready so that, after the publication of the experimental observations, Levine & Steinhardt (1984) were able to make the identification with the Penrose tiling and to call the material 'quasicrystals'.

It has been shown that the geometry of icosahedral quasicrystals can be handled as the bounded projection (slice of finite thickness) through a 6D lattice (Kramer & Neri 1984). This connection with N-dimensional geometry has provided little understanding of the physical basis of quasicrystals which resides in the compromise between local order and global order. Although the Penrose tiling gives the right kind of diffraction pattern, the alloy knows nothing about Penrose's rules of construction and the structure must arise from local interatomic forces.

Though quasicrystals do not have translational periodicity, they possess 'quasi-periodic' translational symmetry. This idea of quasiperiodicity is best illustrated by the Fibonacci sequence in 1D, that was introduced in Section 2.8. A metallurgical example of such a 1D quasicrystal in an Al–Cu–Ni alloy was first pointed out by Chattopadhyay *et al.* (1987).

A significant development from the tiling by two kinds of tiles, regarded as pseudo-unit cells, acute and obtuse rhombuses or rhombohedra, has been to use only *one* kind of packing unit, in the 2D case, a decagon (Gummelt 1995) and in the 3D case, a rhombic triacontahedron (Lord *et al.* 2000; 2001), but allowing the tiles to overlap in certain definite ways. This implies that there is a necessary internal structure and the necessary coincidence of this with that of an overlapping tile provides limitations on their packing to fill space. Steinhardt and others have developed Petra Gummelt's patterns to obtain plausible models for decagonal quasicrystals (Steinhardt *et al.* 1998; Abe *et al.* 2000; Lord & Ranganathan 2001a). These models emphasise local order and indicate how global orientational order can develop from local order.

High-resolution electron micrographs of quasicrystalline phases (Figure 10.6) can be used to identify motifs that make up the structure (Hiraga *et al.* 1985). Quasicrystals have been the most intensely studied intermetallics over the past two decades. All the experimental techniques used for characterising periodic crystals are also suitable for the characterisation of quasicrystals. While the early emphasis was on electron diffraction and microscopy, the availability of large-size single quasicrystals enabled the application of sophisticated X-ray and neutron diffraction methods. Nevertheless the question of 'Where are the atoms in quasicrystals?' has not received a final and convincing answer as yet.

A notable advance came with the realisation that some quasicrystals are stable and find a place in the equilibrium phase diagrams. Al–Pd–Mn, Al–Cu–Fe and

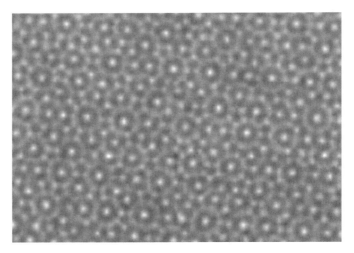

Figure 10.6 High-resolution micrograph from an icosahedral quasicrystal showing fivefold symmetry (courtesy of K. Hiraga).

Al–Ni–Co provide some examples. Thus, it is possible to produce, by slow solidification, single phase quasicrystals of high perfection and purity with dimensions of the order of several centimetres. The original discovery of quasicrystals was made in aluminium–transition metal (Cr, Fe, Mn) alloys. This was rapidly expanded to include several ternary and quaternary alloys. Since then many other systems based on Li, Ga, Mg, Cd, Ti, Zr and Hf have been shown to produce quasicrystals. These may be broadly classified into metastable and stable quasicrystals. Icosahedral quasicrystals can belong to the 'Mackay class' or the 'Bergman class' according to the value of the ratio a_R/d (where a_R is the quasilattice constant and d the average atomic diameter). The ratio is $\sim$1.65 for the former class and $\sim$1.75 for the latter.

The advent of these phases has led to a paradigm shift in our understanding of the atomic configuration in the solid state. As a result of these investigations the definition of crystals underwent a radical change. In 1991 the International Union of Crystallography decided to redefine the term 'crystal' to mean any solid having an essentially discrete diffraction pattern. Within the family of crystals, distinction is made between *periodic crystals*, which are periodic on the atomic scale, and *aperiodic crystals*, which are not.

The growth of mineralogy was helped by the observations of the varied hues and different shapes of minerals. The foundations of crystallography were truly laid by the observations of external form and relating it to the inner atomic structure. The habits of crystals have been extensively studied. Thus it is natural to ask how quasicrystals grow and what form they take. The external morphology of a crystal reflects the internal point group symmetry of the lattice. This aspect is retained in the case of quasicrystals as well. The icosahedral phase grows with external shape of a dodecahedron (Figure 10.7), a rhombic triacontahedron or any of the truncated solids between the two. The internal symmetry is also reflected in the shape of the dendrites.

Figure 10.7 Dodecahedral growth morphology of Al–Cu–Fe quasicrystal.

A major advance occurred when it was realised that quasicrystals are also governed by valence electron concentrations and are essentially Hume-Rothery phases. The Mackay class quasicrystals form at *e/a* of 1.75, while the Bergman class forms at *e/a* of 2.1. The new Cd and Zn base quasicrystals form at *e/a* of 2. This rule has helped in the discovery of new systems forming quasicrystals (Tsai *et al.* 1990).

Over the past 30 years there has been an explosion of publications amounting to more than ten thousand papers. The number of systems displaying quasicrystals has expanded to a more than a hundred. A variety of different types of quasicrystals have been discovered.

10.7 Structurally Complex Intermetallics

In the same paper that he reported on the structure of Mg_2Sn, Pauling also mentioned a study of the crystal structure of $NaCd_2$ (Pauling 1923). The simple and similar stoichiometry of these two compounds is actually deceptive. The X-ray patterns were so complicated that they could not be solved at that time. Not until 1955 was Pauling able to show that $NaCd_2$ is cubic with a lattice parameter of 3 nm and with over 1152 atoms in the unit cell (Pauling 1955). This compound is a fine example of a structurally complex intermetallic. The study of quasicrystals has led to a revival of interest in the elucidation of structure in complex intermetallic compounds. The field is vast and growing. In this section a few examples will be discussed to give a flavour of the subject. These include the Frank–Kasper phases including the Bergman compound (Frank & Kasper 1958), the γ-brass structure, the Mackay rational approximants and Frank's cubic hexagonal phase in a cluster compound.

In 1958, Frank and Kasper identified many compounds with high CNs like 12, 14, 15 and 16 (Figure 3.19). The coordination of 12 corresponds to an icosahedral environment around the atom. In the Laves phase $MgCu_2$, the Mg atoms have a CN of 16 (Figure 3.17), while Cu atoms have icosahedral coordination.

Special interest attaches to the cluster that occurs in $Mg_{32}(Zn, Al)_{49}$ (Bergman *et al.* 1952, 1957) and in the R-phase Al_5CuLi_3 (Audier *et al.* 1988, 1989) described in Section 6.2. The cluster can be visualised in terms of a set of concentric polyhedral 'shells' – an icosahedron, a dodecahedron, a second icosahedron and a truncated icosidodecahedron. The final shell with 60 atoms has the same appearance as the soccer ball with 60 atoms, so familiar in the structure of fullerenes. This 104-atom cluster, Figure 6.10, is an important building block in quasicrystals based on s–p elements such as lithium and magnesium. The quasicrystalline T2 phase is closely related to the crystalline R-phase (Audier *et al.* 1988; Lord *et al.* 2000; 2001). This observation of an icosahedral coordination in a crystalline compound that led Ramachandra Rao & Sastry (1985) to delineate a rational basis for the synthesis of quasicrystals. Kuo and coworkers (Zhang *et al.*

1985) independently used similar reasoning to produce the first titanium based quasicrystal with similarity to the cubic Ti_2Ni compound.

While τ is an irrational number, it can be approximated by a ratio of consecutive Fibonacci numbers (1/1, 2/1, 3/2, etc.). These ratios may be termed *rational approximants* to the irrational golden mean. This concept from number theory can be carried over to the study of quasicrystals, where a series of crystals with large unit cells can be regarded as rational approximants to the quasicrystals. Electron diffraction patterns from these rational approximants are periodic, but the intensity distribution is such that the strong spots mimic the quasicrystalline symmetry. For icosahedral quasicrystals two structures are of importance. Crystalline α-Al–Mn–Si contains 55-atom clusters with three concentric shells – an icosahedron, an icosidodecahedron and a second icosahedron – around a central atom. This has come to be known as the *Mackay icosahedron* (Mackay 1962) and is the important building block in aluminium–transition metal quasicrystals (Figure 6.16). The second structure is the series of compounds based on the Bergman cluster. Tamura (1997) has surveyed a large number of complex intermetallics and drawn attention to the importance of atomic size in this context. An important contribution in this area came from Kreiner & Franzen (1995). They considered the different ways icosahedra can link together (see Figure 6.4) and, in particular, identified an *i*3 cluster (Figure 6.3) in which three icosahedra come together, as an important building unit.

It is evident that several intermetallic compounds with complicated crystal structures, including the examples given above, share the common feature that they may be derived by a decoration of the bcc lattice by clusters of metal atoms. While many authors have tended to emphasise the cluster description, in reality the bonding between clusters is commensurate with bonding within clusters, so the structure is really a 3D network.

In a remarkable coincidence of discovery, three papers appeared in a 1981 issue of *Acta Crystallographica*. These papers show the great affinity in structures of several intermetallics, α-manganese, γ-brass and Ti_2Ni. Though the clusters contain from 22 to 29 atoms, they all have a preponderance of icosahedral coordination (Chabot *et al.* 1981; Hellner & Koch 1981; Nyman & Hyde 1981).

In 1965, Frank introduced the novel idea of projection from a 4D cube to recover the points of a hexagonal lattice with a special c/a ratio of $\sqrt{(3/2)}$. He called it a *cubic hexagonal* crystal, as there was a similarity to the conventional cubic crystals in that directions were perpendicular to planes with the same Miller-Bravais indices (Frank 1965). Other hexagonal lattices with varying c/a ratio can be generated by an affine deformation of the cubic hexagonal lattice.

The lambda phase λ-Al_4Mn is a complex intermetallic compound with a trigonal structure and with 568 atoms in the unit cell. It was solved by Kreiner & Franzen (1995). Ranganathan *et al.* (2002) identified μ-Al_4Mn, λ-Al_4Cr, Mg–Zn–Sm and a host of related intermetallics featuring a special aggregate of icosahedra as Frank's cubic hexagonal phase. The hexagonal phases in the Al–Mn, Al–Cr and Mg–Zn–Re systems have been related to quasicrystals. The metric is generated by the Friauf polyhedra and the icosahedral linkages and leads to a multimetric crystal and there are several interesting connections with hexagonal quasicrystals, hexagonal phases as well as derivative orthorhombic and structures of lower symmetry.

10.8 Metallic Glasses

Turnbull's remarkable demonstration that mercury could be undercooled to as much as $0.67T_m$ (where T_m is melting point temperature) (Turnbull 1952) led Frank (1952) to advance the seminal idea that liquids are characterised by icosahedral coordination, preventing them from easy crystallisation into close-packed structures. Indeed it is this line of thought that led Frank, in collaboration with Kasper (Frank & Kasper 1958), to describe the structure of complex intermetallic compounds in terms of icosahedra and higher coordination polyhedra. In turn, this inspired Bernal (1959) to advance a model for the structure of liquids based on random close-packed structures. This concatenation of ideas over a decade has had an enormous impact on our understanding of metallic structures.

The promising attributes of metallic glasses have led metallurgists to dream of new alloys that would form glasses at low cooling rates like oxide glasses, thus enabling their production in bulk form. This dream was turned into a reality by the pioneering investigations of Inoue starting from 1988. The contributions from his group at Sendai, Japan, documented in a large number of papers in the literature, cover a wide variety of multicomponent alloys and deals with practically all aspects of glass science, and dominates the field (Inoue 1998; 2000). Several significant contributions have come from Johnson's group at Caltech (Johnson 1999), where the pioneering investigations of Duwez gave birth to the field of metallic glasses. While the initial alloys were based on lanthanides, there was a gradual expansion in the elemental basis such that today more than a dozen different metals form the basis for bulk metallic glasses (BMG). They offer for the first time the possibility of studying properties of the glassy and the supercooled liquid states on a temporal and spatial scale previously considered unattainable. A totally unexpected result was that they produce nanocrystals on devitrification, paving the way for bulk nanostructured materials. Some give rise to quasicrystals, leading to bulk nanoquasicrystalline materials and indicating a link with icosahedral order for at least some of these glasses.

The development of a predictive model that would identify the alloy families and the composition ranges that would form bulk glasses remains a grand scientific challenge. Alloy discovery progresses currently by empiricism. The three major contributing factors that facilitate glass formation are atomic size mismatch, high negative heat of mixing and a multicomponent alloy system. These widely accepted empirical rules for the formation of glasses were enunciated by Inoue (1995). While the first two criteria were appreciated from the beginning of the study of metallic glasses, the need for multicomponents has been recognised as of great importance only after the advent of BMG and is often termed as the 'confusion principle' in glass formation. The rationale behind this principle rests on the following arguments. An exploration into the crystal structures of compounds and solid solution phases reveals that as the order of a structure goes from elemental to binary, ternary and higher order, the number of new structure types diminishes. Thus for crystal nucleation to occur in the melt of a multicomponent system, long-range diffusion is necessary. The introduction of a new component in an alloy also introduces local atomic level strain along with chemical disorder, frustrating the crystal. For two atoms of differing sizes, there exists a critical maximum solubility in both the terminal solid solutions. Outside these critical concentrations, the solid solutions become topologically unstable with respect to transformation to the glassy phase. Multicomponent glass-forming alloys contain elements with significant difference in their atomic sizes. With a large difference in atomic sizes the random packing density of the alloy increases.

There is no description for metallic glasses based on random networks, like that in the theory of silica glasses. Instead, the preferred model that has found most favour in describing the metallic glass structure is the polytetrahedral model. The triangle with three spheres at the vertices represents a close-packed configuration. Adding a fourth sphere produces a regular tetrahedral arrangement. Regular tetrahedra cannot pack space, inevitably other configurations arise as more spheres are packed together. There are just eight convex polyhedra with equilateral triangle faces, the *Bernal deltahedra* (Figure 4.9). Some of these can represent voids that can be occupied by metalloid atoms. For example in Pd-base glasses with P as an alloying element, trigonal prisms of Pd atoms surround P atoms. It is also noteworthy that the icosahedron is one of the deltahedra. This appears to be a dominant feature in metal–metal glasses and in particular those based on Zr, Hf and Ti. The crystallisation of some glasses leads to quasicrystals, indicative of the icosahedral order in the glass structure as well. It should be noted, however, that the Bernal-type polytetrahedral structure cannot be an adequate model for metallic glasses; the voids are too sparse to account for the typical compositions of metallic glasses.

If the various triangulated polyhedra are considered, three Bernal (8, 9, 10 vertices) and four Frank-Kasper polyhedra (12, 14, 15 and 16 vertices) appear significant. The absence of polyhedra with 11 and 13 vertices may be noted. If one considers the disclination picture, the Bernal deltahedra are characterised by positive disclinations while the Frank–Kasper polyhedra are marked by negative disclinations. It is interesting to note that, though the metal–metalloid and metal–metal glasses appear to be very different, they are structurally related through this picture of disclinations.

A very promising approach to understanding the structure of metallic glasses is Dan Miracle's cluster-packing model (Miracle 2004) referred to at the end of Section 4.9. The model can accommodate up to three species of 'solute' atoms (α, β and γ) and a 'solvent' atom Ω. For each type of solvent atom there is a characteristic number of Ω atoms that can be accommodated in its first coordination shell – determined by the ratio of atomic radii. In the picture introduced by Miracle the α-clusters are close packed in an fcc array, sharing Ω atoms. The β- and γ-atoms occupy the octahedral and tetrahedral voids. The amorphous nature of the material thus resides principally in the randomness of the resulting positioning of the solute atoms. The model is able to predict metallic glass compositions quite accurately.

Employing sophisticated X-ray experimental techniques, Sheng *et al.* (2006) have obtained detailed information about the short- and medium-range structure of a number of binary metallic glasses. This work reveals the range and statistical distribution of Voronoi regions and related coordination shells of solute atoms, and the influence of atomic size ratio on this distribution. Frank-Kaspar coordination polyhedra predominate. Medium-range patterns involving clustering of the basic cluster types is also revealed. This work marks a very important advance in the understanding of the metallic glass state, providing a test for hypothetical models such as those proposed by Miracle (2004).

10.9 Nanocrystals

Richard Feynman pointed out as early as 1959 that there is 'room at the bottom' and in a visionary statement outlined the possibilities offered by nanostructured materials (Feynman 1960). The next major event came in 1984, when Herbert Gleiter synthesised nanocrystals by inert gas condensation and gave a name and habitation to nanostructured materials. Nanostructured materials are defined as those materials with at least one dimension less than 100 nm (Gleiter 2000). The volume fraction of atoms at the interfaces exceeds at least 1%. Figure 10.8 shows a TEM of nanocrystalline palladium (Ranganathan *et al.* 2001). The preponderance of atoms at the interfaces is expected to confer certain unique properties on these materials.

The initial studies yielded nanometals in small quantities and progress was slow. But a remarkable recent development has been the preparation of bulk nanocrystalline materials by devitrification or direct quenching from the liquid. This can arise due to phase separation in the liquid or glass. Such a fine-scale separation preceding crystallisation determines the scale of crystallisation to be nanometric. On

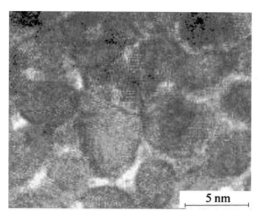

5 nm

Figure 10.8 TEM of nanocrystalline palladium.

the other hand a large number of quenched-in nuclei could lead to a high nucleation rate and slow growth. Inoue (2000) prescribed that such nanocrystallisation in Zr alloys is promoted by the addition of elements such as Au, Ag, Pd and Pt. These have strong interactions with Zr and could lead to clustering in the glass and consequent nanocrystallisation. In particular, in several glasses it leads to nanoquasicrystals. This is an active area of research (Li *et al.* 2001; Louzguine *et al.* 2001).

10.10 Helices

As we saw in Chapter 7, there is an intimate connection between space filling and helices. Helices and dense packing of spherical objects are two closely related problems (Boerdijk 1952; Lord & Ranganathan 2001b; Lord 2002). For instance, the Boerdijk–Coxeter helix, which is obtained as a linear packing of regular tetrahedra, is a very efficient solution to some close-packing problems. The shapes of biological helices result from various kinds of interaction forces, including steric repulsion. Thus, the search for maximum density can lead to structures related to the Boerdijk–Coxeter helix. Sadoc and Rivier have related the geometry of the Boerdijk–Coxeter helix to the α-helix structure in proteins and on this basis have modelled the intricate triple-helix structure of the protein collagen (Sadoc & Rivier 1999; Lord & Ranganathan 2001b). The characteristic aperiodicity of the helices permits a description of these molecules as 'biological quasicrystals'. This geometrical approach to protein structure initiated by Sadoc & Rivier (1999a, b) and Sadoc & Mosseri (1999), is based on the projection from the 4D polytope {3, 3, 5} (Sadoc 2001). Particularly striking examples of helices in highly complex metallics are the close-packed tetrahelical rods in β-manganese (Figure 7.7) and the amazing double helix of icosahedra discovered by Boström & Lidin (2002) in a cobalt–zinc phase (Figure 7.8).

10.11 Clathrates

The duals of the Frank–Kasper polyhedral space fillings, obtained by joining centres of every pair of tetrahedra in face contact, were met in Chapter 3. They are space fillings by polyhedra with 12, 14 , 15 and 16 faces. Figure 3.23 shows the hydrogen bond framework of chlorine hydrate, a structure describable as a space filling of pentagonal dodecahedra and 14-hedra. A second clathrate contains dodecahedra and 16-hedra (Figure 3.22) and is a dual of the $MgCu_2$ structure discussed above. There is now a resurgence of interest in the clathrates among metallurgists, as silicon and germanium form such structures with the cages occupied by alkali atoms.

A minimal area foam with the smallest area per unit volume has attracted the attention of metallurgists because of its relevance for metallic grain shapes. For a long time the solution by Kelvin was taken to be the optimum until the discovery of Weaire & Phelan (1994) that a packing of dodecahedra and 14-hedra gives a slightly smaller area per unit volume. Thus we find the same considerations in such diverse materials as intermetallics, clathrates and soap foams.

10.12 Conclusion

The objective of this chapter has been to draw attention to the diversity of structures offered by metallics, many of which have been uncovered in recent years, and thus to draw attention to the way that the geometrical principles presented in earlier chapters provide tools for elucidating these structures. The same geometrical principles that underlie the structures of alloys – which occur in a bewildering variety of intricate phases – also govern the constitution of naturally occurring minerals and the complex molecular architecture of biological systems. These interconnected geometrical principles arise from a study of the kinds of forms that are possible in 3D-Euclidean space, and form the basis of a 'science of morphology'.

Appendix: Handling the New Geometry

What is usually required as the output from investigations of the geometry related to new materials is pictures in two dimensions (2D) or solid models to be produced by a stereolithography process, plus calculations of various numerical parameters such as coordinates, areas or volumes or curvatures. Computer systems are now always in a fluid state with new developments, both in software and hardware, and with familiar systems becoming obsolete. However, we must mention briefly some of the current systems which we have used and which we recommend.

In the first place we must put the commercial software, **Mathematica**, now at Version 5.1, which leads in the field of general computational software. It is immensely complex and is the highest level language available, but one can begin in the simplest way with $3 + 4 = 7$ and move gradually to the most sophisticated mathematics. The creators of this high level language chose, for example, to use complete recognisable words for the various processes so that it is not necessary to remember lists of arbitrary abbreviations. The system is interpreted, each statement being executed in turn, but within a single statement complex procedures are compiled. It was influenced by the original Dartmouth College **BASIC**, developed from 1962 by John Kemeny and Thomas Kurtz, with the aim of making computing available, not just to science students, but to all students. In particular, the system was not patented or copyrighted but was free to all. Mathematica was also influenced by the language APL developed by Iverson about 1962 which enabled the manipulation of blocks of data as single symbols.

A major feature of the original BASIC was the provision for solving a matrix of linear equations where complexity was concealed in a function INV[ ] which inverted a matrix. This was not implemented in most of the subsequent version of BASIC which lost its initial simplicity for matrix operations.

Mathematica produces almost all conceivable types of output: numerical, graphical, algebraic and textual with great power and simplicity. For example, a 3D function can be plotted and then can be output as an *.STL file appropriate for stereolithography.

Nevertheless, we must mention **PowerBasic**, (now Version 3.5 for DOS and 8.01 for Windows) which is a commercial package, which has a very simple user interface and gives the user the speed and power of the underlying C++ language. PowerBasic is compiled and runs extremely fast. Programs written in Fortran can be readily transcribed into PowerBasic.

We have used two freely available software systems which merit attention.

The **Surface Evolver** system, produced and maintained by Ken Brakke, is basically a finite element analysis system which solves all problems to do with surfaces conditioned by all kinds of forces and constraints, besides surface tension. We have found that it has still wider applications. It can produce pictures but also files of 3D structure (*.QUAD) which enable structures to be transferred to other applications.

For the production of realistic pictures the free system **Povray** is invaluable. This comprises point-of-vision ray tracing and will provide pictures of scenes, perhaps of artistic merit. Its use will absorb whatever time the user may have available, but the results can be very impressive.

Besides the standard graphics systems such as **CorelDRAW, Paint Shop Pro** and **Photoshop**, which are very convenient for producing pictures, there are special programs for handling crystal structures. In particular **Mercury**, freely available, handles structures from the Cambridge Crystallographic Databank, the formats of which are standardised as *.CIF and *PDB, files produced by crystallographers and those from the protein databases.

SolidView is freely available and enables *STL files to be visualised.

For general text and manuscript handling we recommend **Acrobat Reader** (now Version 6.0) which is free and also **OpenOffice**, which is a large free system replicating commercial software for text, standard slide show presentations and such applications.

References

Abe, E., Saitoh, K., Takakura, H., Tsai, A. P., Steinhardt, P. J. & Jeong, H.-C. Quasi-unit-cell model for an Al-Ni-Co ideal quasicrystal based on clusters with broken tenfold symmetry. *Phys. Rev. Lett.* 84 (2000) 4609–12.

Almgren Jr, F. J. *Plateau's Problem: an Introduction to Varifold Geometry.* Benjamin, New York (1966); revised edition, American Mathematical Society, Providence, RI (2001).

Amelinckx, S., Zhang, X. B., Bernaerts, D., Zhang, X. F., Ivanov, V. & Nagy J. B. A formation mechanism for catalytically grown helix shaped graphite nanotubes. *Science* 265 (1994) 635–9.

Andersson, S. An alternative description of the structure of Rh_7Mg_{44} and Mg_6Pd. *Acta Cryst. A* 34 (1978) 833–5.

Andersson, S. & Hyde, S. T. The intrinsic curvature of solids. *Z. Kristallogr.* 168 (1984) 1–17.

Andreini, A. Sulle reti di poliedri regolare e semiregolari e sulle correspondenti reti correlative. *Memorie della Societa Italiane delle Scienze* 14 (1907) 75–129.

Applebaum, J. & Weiss, Y. The packing of circles on a hemisphere. *Meas. Sci. Technol.* 10 (1999) 1015–19.

Aste, T. Circle, sphere and drop packings. *Phys. Rev. B* 53 (1996) 2571–9.

Aste, T. & Weaire, D. *The Pursuit of Perfect Packing.* Institute of Physics Publishing, Bristol, UK (2000).

Aston, M. W. Icosahedral packing of equal spheres using five basic cells. *Hyperspace* 8(2) (1999) 34–52.

Atiyah, M. & Sutcliffe, P. Polyhedra in physics, chemistry and geometry. *Milan J. Math.* 71 (2003) 33–58.

Audier, M. & Duneau, M. Icosahedral quasiperiodic packing of fibres parallel to fivefold and threefold axes. *Acta Cryst. A* 56 (2000) 49–61.

Audier, M. & Guyot, P. Al_4Mn quasicrystal atomic structure, diffraction data and Penrose tiling. *Phil. Mag. B* 3 (1986) L43–51.

Audier, M. & Guyot, P. The structure of the icosahedral phase – atomic decoration of the basic cells. In: *Quasicrystalline Materials: Proc. ILL/Codest Workshop, Grenoble 1988* (Janot, Ch. & Dubois, J. M., eds). World Scientific, Singapore (1988) 181–94.

Audier, M., Pannetier, J., Leblanc, M., Janot, C., Lang, J. M. & Dubost, B. An approach to the structure of quasicrystals: a single crystal X-Ray and neutron diffraction study of the R-Al_5CuLi_3 phase. *Physica B* 153 (1988) 136–42.

Audier, M., Janot, Ch., de Boissieu, M. & Dubost, B. Structural relationships in intermetallic compounds of the Al-Li-(Cu, Mg, Zn) system. *Phil. Mag. B* 60 (1989) 437–88.

Baer, S. *Zome Primer.* Zomeworks Corpn., Albuquerque, NM (1970).

Baer, S. The 31-zone structural system. In: *Third International Conference on Space Structures* (Nooshin, H., ed.). Elsevier, New York (1984) 872–5.

Baerlocher, Ch., Meier, W. M. & Olson, D. H. *Atlas of Zeolite Framework Types* (5th ed.). Elsevier, New York (2001).

Bausch, A. R., Bowick, M. J., Cacciuto, A., Dinsmore, A. D., Hsu, M. F., Nelson, D. R., Nikolaides, M. G., Travesset, A. & Weitz, D. A. Grain boundary scars and spherical crystallography. *Science* 299 (2003) 1716–8.

Belin, C. H. E. & Belin, R. C. H. Synthesis and crystal structure determinations in the Γ and δ phase domains of the iron-zinc system: electronic and bonding analysis of $Fe_{13}Zn_{39}$ and $FeZn_{10}$, a subtle deviation from the Hume-Rothery standard? *J. Solid state Chem.* 151 (2000) 85–95.

Bergman, G., Waugh, J. L. T. & Pauling, L. Crystal structure of the intermetallic compound $Mg_{32}(Al, Zn)_{49}$ and related phases. *Nature* 169 (1952) 1057–8; The crystal structure of the metallic phase $Mg_{32}(Al, Zn)_{49}$, *Acta Cryst.* 10 (1957) 254–9.

Bernal, J. D. A geometrical approach to the structure of monatomic liquids. *Nature* 183 (1959) 141–7.

Bernal, J. D. Geometry of the structure of monatomic liquids. *Nature* 185 (1960a) 68–70.

Bernal, J. D. The structure of liquids. *Sci. Am.* 201 (1960b) 124–31.

Bernal, J. D. The structure of liquids. *Proc. Roy. Soc. London A* 208 (1964a) 299–322.

Bernal, J. D. The structure of liquids. *New Sci.* 8 (1964b) 453–5.

Bernal, J. D. & Carlisle, C. H. The range of generalised crystallography. *Kristallografiya* 13 (1968) 927–951; *Sov. Phys. – Cryst.* 13 (1969) 811–31.

Bernal, J. D., Cherry, I. A., Finney, J. L. & Knight, K. R. An optical machine for measuring sphere coordintes in random packings. *J. Phys. E: Sci. Instrum.* 3 (1970) 388–90.

Bilinski, S. Über die Rhombenisoeder. *Glasnik matematicko-fisicki i astronomeski* 16 (1960) 251–62.

Blatov, V. A. & Shevchenko, A. P. Analysis of voids in crystal structures: the methods of 'dual' crystal chemistry. *Acta Cryst. A* 59 (2003) 34–44.

Blum, Z., Lidin, S. & Thomasson, R. Zeolites: conveyors of non-Euclidean geometry. *Angew. Chem. Int. Ed. Engl.* 27 (1988) 953–6.

Boerdijk, A. H. Some remarks concerning close-packing of equal spheres. *Philips Res. Rep.* 7 (1952) 303–13.

Bonnet, O. Note sur la théorie générale des surfaces. *Comptes Rendu* 37 (1853) 529–32.

Booth, D. The new Zome primer. In: *Fivefold Symmetry* (Hargittai, I., ed.). World Scientific, Singapore (1992) 221–33.

Boström, M. & Lidin, S. Preparation and double-helix icosahedra structure of δ-Co_2Zn_{15}. *J. Solid State Chem.* 166 (2002) 53–7.

Bourgoin, J. *Les Eléments de l'Art Arabe: le Trait des Entrelacs.* Firmin-Didot, Paris (1879); republished as *Arabic Geometrical Pattern and Design.* Dover, New York (1973).

Bradley A. J. & Jones, P. An X-ray investigation of the copper-aluminium alloys. *J. Inst. Met.* 51 (1933) 131–62.

Bradley, A. J & Thewlis, J. The structure of γ-brass. *Proc. Roy. Soc. London A* 112 (1926) 678–81.

Brakke, K. A. The Surface Evolver. *Experimental Math.* 1 (1992) 141–165.

Brakke, K. A. The Surface Evolver and the stability of liquid surfaces. *Phil. Trans. Roy. Soc. London A* 354 (1996) 2143–57.

Brakke, K. A. *Plateau's Problem.* American Mathematical Society Providence, RI (2001).

Bravais, L. & Bravais A. Essai sur la disposition des feuilles curvisériées. *Ann. Sci. Nat. Bot. Biol. Veg.* 7 (1837) 42–110, 193–221, 291–348; 8 (1838) 11–42.

Brunner, G. O. Parameters for frameworks and space filling polyhedra. *Z. Kristallogr.* 156 (1981) 295–303.

Buckminster Fuller, R. *Synergetics: Exploration in the Geometry of Thinking.* Macmillan, New York (1975).

Catalan, E.C. Mémoires sur la theorie des polyèdres, *J. École Polytech.* 24 (1865) 1–71.

Chabot, B., Cenzual, K. & Parthé, E. Nested polyhedra units: a geometrical concept for describing cubic structures. *Acta Cryst. A* 37 (1981) 6–11.

Charvolin, J. & Sadoc, J.-F. Ordered bicontinuous films of amphiphiles and biological membranes. *Phil. Trans. Roy. Soc. London A* 354 (1996) 2173–92.

Chattopadhyay, K., Lele, S., Prasad, R., Ranganathan, S., Subbanna, G. N. & Thangaraj, N. On the variety of electron diffraction patterns from quasicrystals. *Scripta Metall.* 19 (1985a) 1331–4.

Chattopadhyay, K., Lele, S., Ranganathan, S., Subbanna G. N. & Thangaraj, N. Electron microscopy of quasicrystals and related structures. *Curr. Sci.* 54 (1985b) 895.

Chattopadhyay, K., Lele, S., Thangaraj, N. & Ranganathan, S. Vacancy ordered phases and one dimensional quasiperiodicity. *Acta Metall.* 35 (1987) 727–33.

Chen, B., Eddaoudi, M., Hyde, S. T., O'Keeffe, M. & Yaghi, O. M. Intewoven metal-organic framework on a periodic minimal surface with extra large pores. *Science* 291 (2001) 1021–23.

Chorbachi, W. K. In the tower of Babel: beyond symmetry in Islamic design. *Comp. Maths with Appl.* 17 (1989) 751–89.

Chung, F. & Sternberg, S. Mathematics and the Buckyball. *Am Scientist* 81 (1993) 58–71.

Church, A. H. *The Relation of Phyllotaxis to Mechanical Laws.* Williams and Norgate, London (1904).

Clare, B. W. & Kepert, D. L. The closest packing of equal circles on a sphere. *Proc. Roy. Soc. London A* 405 (1986) 329–44.

Clare, B. W. & Kepert, D. L. The optimal packing of circles on a sphere, *J. Math. Chem.* 6 (1991) 325–49.

Cockayne, E. & Widom, M. Structure and phason energetics of Al-Co decagonal phases. *Phil. Mag. A* 77 (1998) 593–619.

Conway, J. H. The orbifold notation for surface groups. In: *Groups, Combinatorics and Geometry* (Liebeck, M. & Sakl, J., eds). Cambridge University Press (1992) 438–47.

Conway, J. H. & Huson, D. H. The orbifold notation for two-dimensional groups. *Struc. Chem.* 13 (2002) 247–58.

Conway, J. H. & Knowles, K. M. Quasiperiodic tiling in two and three dimensions. *J. Phys. A: Math. Gen.* 19 (1986) 3645–53.

Conway, J. H. & Sloane, N. J. A. *Sphere Packings, Lattices and Groups.* Springer, New York (1988); 2nd edn. (1998).

Cook, T. A. *The Curves of Life, Being an Account of Spiral Formations and Their Application to Growth in Nature, To Science and to Art.* Constable London (1914); Dover, New York (1979).

Cornelli, A., Farinato, R. & Loreto, L. Environments of points and related polyhedral configurations: an interactive computer graphics (ICG) approach. *Per. Mineral. Roma* 53 (1984) 135–58.

Costa, C. Imersões minimas completas em R^3 de genero un e curvatura total finita. *Doctoral Thesis, IMPA, Rio de Janeiro, Brasil* (1982); Example of a complete minimal immersion in R^3 of genus one and three embedded ends. *Bull. Soc. Bras. Mat.* 15 (1984) 47–54.

Cottrell, A. H. *Concepts in the Electron Theory of Metals.* Institute of Materials UK, London (1998).

Coxeter, H. S. M. Regular skew polyhedra in three and four dimensions and their topological analogues. *Proc. London Math. Soc. (Ser. 2)* 43 (1937) 33–62.

Coxeter, H. S. M. *Regular Polytopes.* Macmillan, New York (1963); Dover, New York (1973).

Coxeter, H. S. M. *Regular Complex Polytopes*. Cambridge University Press (1974).

Coxeter, H. S. M. *Introduction to Geometry*. Wiley, New York (1969); 2nd edn (1989).

Coxeter, H. S. M. The simplicial helix and the equation tan $n\theta$ = n tan θ. *Canad. Math. Bull.* 28 (1985) 385–93.

Coxeter, H. S. M. A simple introduction to colored symmetry. *Int. J. Quantum Chem.* 31 (1987) 455–61.

Coxeter, H. S. M. & Moser, W. O. J. *Generators and Relations for Discrete Groups*. Springer, New York, Berlin (1957).

Coxeter, H. S. M., Longuet-Higgins, M. S. & Miller, J. C. P. Uniform polyhedra. *Philos. Trans. Roy. Soc. London A* 246 (1953) 401–50.

Critchlow, K. *Time Stands Still*. Gordon Fraser London (1979).

Critchlow, K. & Nasr, S. H. *Islamic Patterns: An Analytical and Cosmological Approach*. Thames and Hudson London (1979); Inner Traditions Intl. Ltd., Rochester, VT (1999).

Critchlow, K. *Order in Space: A Design Sourcebook*. Thames and Hudson (1969); 2nd edn (2000).

Cromwell, P. R. Kepler's work on polyhedra. *Math. Intelligencer* 17 (1995) 23–33.

Cromwell, P. R. *Polyhedra*. Cambridge University Press (1999).

Crowe, D. W. The mosaic patterns of H. J. Woods. *Comp. & Maths with Appl.* 12B (1986) 407–11.

Cundy, H. M. & Rollett, A. P. *Mathematical Models*. Oxford University Press (1951). 2nd edn (1961).

Curl, R. F. & Smalley, R. E. Fullerenes. *Sci. Am.* 265 (1991) 54–63.

Davis, M. E. Ordered porous materials for emerging applications. *Nature* 417 (2002) 813–20.

De Bruijn, N. G. Algebraic theory of Penrose's non-periodic tilings of the plane. *Math. Proc. A* 84 (1981) 39–66.

Delgado-Friedrichs, O. & Huson, D. A combinatorial theory of tilings. In: *Voronoï's Impact on Modern Science. Book II* (Engel, P. & Syta, H., eds). National Academy of Science of Ukraine, Institute of Mathematics, Kiev (1998) 85–95.

Delgado-Friedrichs, O. & Huson, D. H. Orbifold triangulations and crystallographic groups. *Per. Math. Hung.* 34 (1997) 29–55.

Delgado-Friedrichs, O. & Huson, D. H. Tiling space by Platonic solids, I. *Discrete Computational Geom.* 21 (1999) 299–315.

Delgado-Friedrichs, O., Dress, A. W. M., Huson, D. H., Klinowski, J. & Mackay, A. L. Systematic enumeration of crystalline networks. *Nature* 400 (1999a) 644–7.

Delgado-Friedrichs, O., Dress, A. W. M. & Huson, D. H. Tilings and symbols: a report on the uses of symbolic calculation in tiling theory. *Hyperspace* 8(3) (1999b) 10–25.

Delgado-Friedrichs, O., Plévert, J. & O'Keeffe, M. A simple isohedral tiling of three-dimensional space by infinite tiles and with symmetry $I a \bar{3} d$. *Acta Cryst. A* 58 (2002) 77–8.

Delgado-Friedrichs, O., O'Keeffe, M. & Yaghi, O. M. Three-periodic nets and tilings: regular and quasiregular nets. *Acta Cryst. A* 59 (2003) 22–7.

Delgado-Friedrichs, O. & O'Keeffe, M. Isohedral simple tilings: trinodal and by tiles with ≤ 16 faces. *Acta Cryst. A* 61 (2005) 358–62.

Delone, B. N., Padurov, N. & Aleskandrov, A. *Mathematicheskie Osnovy Strukturnogo Analiza Kristallov. [The Mathematical Bases of Crystal Structure Analysis]* ONTI GTTI Leningrad & Moscow (1934).

Delone, B. N., Dolbilin, N. P., Shtogrin, M. J. & Galiulin, R. V. A local criterion for regularity of a system of points. *Sov. Math. Doklady* 17 (1976) 319–32.

Dirichlet, G. L. Über die Reduction der positiven quadratische Formen mit drei unbestimmten ganzen Zählen. *J. reine angew. Math.* 40 (1850) 216–9.

Doye, J. P. K. & Wales, D. J. The structure of (C60)N clusters. *Chem. Phys. Lett.* 262 (1996) 167–74.

Doye, J. P. K. & Wales, D. J. Polytetrahedral clusters. *Phys. Rev. Lett.* 86 (2001) 5719–22.

Dress, A. W. M. Regular polytopes and equivariant tesselations from a combinatorial point of view. In: *Springer Lecture Notes in Mathematics*, vol. 1172. Springer, Göttingen (1985) 56–72.

Dress, A. W. M. Presentations of discrete groups, acting on simply connected manifolds, in terms of parametrized systems of Coxeter matrices: a systematic approach. *Adv. In Math.* 63 (1987) 196–212.

Dress, A. W. M., Huson, D. H. & Molnar, E. The classification of face-transitive three-dimensional tilings. *Acta Cryst. A* 49 (1993) 806–17.

Duneau, M. & Audier, M. Quasiperiodic packing of fibres with icosahedral symmetry. *Acta Cryst. A* 55 (1999) 746–54.

Dunlap, R. A. *The Golden Ratio and Fibonacci Numbers*. World Scientific, Singapore (1998).

Dürer, A. *Unterweysung der Messung*. H. Formschneyder, Nürnberg (1523).

Du Val, P. *Homographies, Quaternions and Rotations*. Oxford University Press (1964).

Dyer, A. *An Introduction to Zeolite Molecular Sieves*. Wiley, New York (1988).

Elam, K. *Geometry of Design: Studies in Proportion and Composition*. Princeton Architectural Press, Princeton, NJ (2001).

El-Said, I. & Parman, A. *Geometrical Concepts in Islamic Art*. World of Islam Festival Publishing Co., London (1976).

Elser, V. Space filling minimal surfaces and sphere packings. *J. Phys. 1 France* 4 (1994) 731–5.

Elser, V. A cubic Archimedean screw. *Phil. Trans. Roy. Soc. London A* 354 (1996) 2071–5.

Elser, V. & Henley, C. L. Crystal and quasicrystal structures in Al-Mn-Si alloys. *Phys. Rev. Lett.* 55 (1985) 2883–6.

Erickson, R. O. Tubular packing of spheres in biological fine structures. *Science* 181 (1973).

Fedorov, E. S. The symmetry of regular systems of figures (in Russian), *Zap. Mineralog. Obsc.* 28(2) (1891) 1–146. English translation in *Symmetry of Crystals. ACA Monograph 7. Amer. Crystallographic Assoc.* New York (1971) 50–131.

Fejes Tóth, L. *Regular Figures*. Macmillan, New York (1964).

Fejes Tóth, L. L. *Lagerungen in der Ebene auf der Kugel und im Raum*. Springer, Berlin (1953); 2nd edn (1972).

Ferey, G. Simplicity of complexity – rational design of giant pores. *Science* 291 (2001) 994–5.

Ferro, A. C. & Fortes, M. A. A new family of trivalent space-filling parallelohedra. *Z. Kristallogr.* 173 (1985) 41–57.

Feynman, R. P. There's plenty of room at the bottom. *Eng. and Sci. (CalTech)* 23 (February 1960) 22–36. Reprinted in *Nanotechnology: Research and Perspectives* (Crandall, B. C. & Lewis, J., eds). MIT Press, Cambridge, MA (1992) 347–63, and in *Miniaturization* (Gilbert, D. H., ed.). Reinhold, New York (1961) 282–96. See also http://www.zyvex.com/nanotech/ feynman.html.

Finney, J. L. Random packings and the structure of simple liquids. I. The geometry of random close packing. *Proc. Roy. Soc. London A* 319 (1970) 479–93.

Finney, J. L. Fine structure in randomly packed, dense clusters of hard spheres. *Mater. Sci. Eng.* 23 (1976) 199–205.

Fischer, W. Tetragonal sphere packings. I. Lattice complexes with zero or one degree of freedom. *Z. Kristallogr.* 194 (1991a) 67–85; II. Lattice complexes with two degrees of freedom. *Z. Kristallogr.* 194 (1991b) 87–110; III. Lattice complexes with three degrees of freedom. *Z. Kristallogr.* 205 (1993) 9–26.

Fischer, W. & Koch, E. On 3-periodic minimal surfaces. *Z. Kristallogr.* 179 (1987) 31–52.

Fischer, W. & Koch, E. New surface patches for minimal balance surfaces. I. Branched catenoids. *Acta Cryst. A* 45 (1989a) 166–169; III. Infinite strips. *Acta Cryst. A* 45 (1989b) 485–90.

Fischer, W. & Koch, E. Genera of minimal balance surfaces. *Acta Cryst. A* 45 (1989c) 726–32.

Fischer, W., Burslaff, H., Hellner, E. & Donnay, J. D. H. *Space Groups and Lattice Complexes*. National Bureau of Standards Monograph 134. US Government printing Office, Washington DC (1973).

Fischer, W. & Koch, E. Lattice complexes. In: *International Tables for Crystallography A* (Hahn, J. T., ed.). Kluver Academic Publishers, Dordrecht, The Netherlands (1995) 825–54.

Fischer, W. & Koch, E. Spanning minimal surfaces. *Phil. Trans. Roy. Soc. London A* 354 (1996a) 2105–42.

Fischer, W. & Koch, E. Two 3-periodic self-intersecting minimal surfaces related to the Cr$_3$Si structure type. *Z. Kristallogr.* 211 (1996b) 1–3.

Fogden, A. Description of a 3-periodic minimal surface family with trigonal symmetry. *Z. Kristallogr.* 209 (1994) 22–31.

Fogden, A. & Haeberlein, M. New families of triply periodic minimal surfaces. *J. Chem. Soc. Faraday Trans.* 90 (1994) 263–70.

Fogden, A. & Hyde, S. T. Parametrization of triply periodic minimal surfaces. I. Mathematical basis of the construction algorithm for the regular class. *Acta Cryst.* 148 (1992a) 442–51; II. Regular class solutions. *Acta Cryst.* 148 (1992b) 575–91; III. General algorithm and specific examples for the irregular class. *Acta Cryst.* 149 (1993) 409–21.

Fogden, A. & Hyde, S. T. Continuous transformations of cubic minimal surfaces. *Eur. Phys. J. B* 7 (1999) 91–104.

Fogden, A., Haeberlein, M. & Lidin, S. Generalizations of the gyroid surface. *J. Phys. 1 France* 3 (1993) 2371–85.

Fowler, P. W. & Tarnai, T. Transition from circle packing to covering on a sphere: the odd case of 13 circles. *Proc. Roy. Soc. London A* 155 (1999) 4131–43.

Frank, F. C. Supercooling of liquids. *Proc. Roy. Soc. London A* 215 (1952) 43–6.

Frank, F. C. On Miller-Bravais indices and four-dimensional vectors. *Acta Cryst.* 18 (1965) 862–6.

Frank, F. C. & Kasper, J. S. Complex alloy structures regarded as sphere packings. I. definitions and basic principles. *Acta Cryst.* 11 (1958) 184–90.

Gardner, M. Extraordinary non-periodic tiling that enriches the theory of tiles. *Sci. Amer.*, January (1977) 110–21.

Gardner, M. *Penrose Tiles to Trapdoor Cyphers*. Freeman, New York (1995).

Ghyka, M. C. *The Geometry of Art and Life*. Sheed & Ward, New York (1946); Dover, New York (1977).

Gips, J. *Shape Grammars*. Birkhäuser, Basel (1975).

Gleiter, H. Nanostructured materials: basic concepts and microstructure. *Acta Mater.* 48 (2000) 1–29.

Goetzke, K. & Klein, H.-J. Properties and efficient algorithmic determination of different classes of rings in finite and infinite polyhedral networks. *J. Non-Crystalline Solids* 127 (1991) 215–20.

Gotoh, K. & Finney, J. L. Statistical geometrical approach to random packing density of equal spheres. *Nature*, 252 (1974) 202–5.

Gozdz, W. & Holyst, R. High genus periodic gyroid surfaces of nonpositive Gaussian curvature. *Phys. Rev. Lett.* 76 (1996a) 2726–9.

Gozdz, W.T. & Holyst, R. Triply periodic surfaces and multiply continuous structures from the Landau model of microemulsions. *Phys. Rev. E* 54 (1996b) 5012–27.

Gozdz, W. & Holyst, R. From the Plateau problem to periodic minimal surfaces in lipids, surfactants and diblock copolymers. *Macromo. Theory Simul.* 5 (1996c) 321–32.

Grünbaum, B. Uniform tilings of 3-space. *Geombinatorics* 4 (1994) 49–56.

Grünbaum, B. & Shephard, G. C. In: *The Mathematical Gardner* (Klarner, D. A., ed.). Wadsworth International, Belmont, CA (1981).

Grünbaum, B. & Shephard, G. C. *Tilings and Patterns*. W. H. Freeman, New York (1987).

Gummelt, P. Construction of Penrose tilings by a single aperiodic proto set. In: *Procceedings of the 5th International Conference on Quasicrystals* (Janot, C. & Mosseri, R., eds). World Scientific, Singapore (1995) 84–7.

Gummelt, P. Penrose tilings as coverings of congruent decagons. *Geometriae Dedicata* 62 (1996) 1–17.

Gummelt, P. *Aperiodische Überdeckungen mit einem Clustertyp*. Thesis, University of Greifswald (1998).

Gummelt, P. & Bandt, C. A cluster approach to random Penrose tilings. *Mater. Sci. Eng. A* 294–296 (2000) 250–3.

Hägg, G. *Zeitschr. f. physik. Chemie*. 12 (1930–31) 33.

Hahn, J. T. (ed.). *International Tables for Crystallography A*. Kluver, Dordrecht (1995).

Hales, T. C. Sphere packings, I. *Discrete Computational Geometry* 17 (1996) 1–51; Sphere packings, II. *Discrete Computational Geometry* 18 (1997) 135–49.

Hamilton, W. R. *Lectures on Quaternions: Containing a Systematic Statement of a New Mathematical Method*. Hodges and Smith, Dublin (1853).

Hamada, N., Sawada, S. & Oshiyama, A. New one-dimensional conductors – graphitic nanotubes. *Phys. Rev. Lett.* 68 (1992) 1579–81.

Hamkins, J. & Zeger, K. Asymptotically dense spherical codes – part I: wrapped spherical codes. *IEEE Trans. Inform. Theory* 43 (1997a) 1774–85.

Hamkins, J. & Zeger, K. Asymptotically dense spherical codes – part II: laminated spherical codes. *IEEE Trans. Inform. Theory* 43 (1997b) 1786–98.

Han, S. & Smith, J. V. Enumeration of four-connected three-dimensional nets. I. Conversion of all edges of simple three-connected two-dimensional nets into crankshaft chains. *Acta Cryst. A* 55 (1999) 332–41; II. Conversion of edges of three-connected 2D nets into zigzag chains. *Acta Cryst. A* 55 (1999) 342–59; III. Conversion of edges of three-connected two-dimensional nets into saw chains. *Acta Cryst. A* 55 (1999) 360–82.

Hargittai, I. (ed.). *Fivefold Symmetry*. World Scientific, Singapore (1992).

Harper, P. E. & Gruner, S. M. Electron density modeling and reconstruction of infinite periodic minimal surfaces (IPMS) based phases in lipid-water systems. I. Modeling IPMS based phases. *Eur. Phys. J. E* 2 (2000) 217–28.

Harrison, W. A. The Fermi surface. *Science* 134 (1961) 915–20.

Hart, G. W. & Picciotto, H. *Zome Geometry: Hands-on Learning with Zome Models*. Key Curriculum Press, Emeryville, CA (2002).

Haüssermann, U., Svensson, C. & Lidin, S. Tetrahedral stars as flexible basis clusters in sp-bonded intermetallic frameworks and the compound $BaLi_7Al_6$ with the $NaZn_{13}$ structure. *J. Am. Chem. Soc.* 120 (1998) 3867–80.

Heesch, H. & Laves, F. Über dünne Kugelpackungen. *Z. Kristallogr.* 85 (1933) 443–53.

Hellner, E. Descriptive symbols for crystal structure types and homeotypes based on lattice complexes. *Acta Cryst.* 19 (1965) 703–12.

Hellner, E. & Koch, E. Cluster or framework considerations for the structures of Tl_7Sb_2, α-Mn, Cu_5Zn_8 and their variants $Li_{22}Si_{51}$, $Cu_{41}Sn_{11}$, $Sm_{11}Cd_{45}$, Mg_6Pd and Na_6Tl with octuple unit cells. *Acta Cryst. A* 37 (1981) 1–6.

Herz-Fischler, R. A. *Mathematical History of the Golden Number*. Dover, New York (1998).

Higgins, J. B. Silica zeolites and clathrates. In: *Silica: Physical Behaviour, Geochemistry and Material Applications*. (Heaney, P. J., Prewitt, C. T. & Gibbs, G. V. eds). Mineralogical Society of America, Chantilly, VA (1994) 508–43.

Hilbert, D. & Cohn-Vossen, S. *Geometry and the Imagination*. Chelsea Publishing Company, London (1952).

Hiraga, K., Hirabayashi, M., Inoue, A. & Masumoto, T. Icosahedral quasicrystals of a melt-quenched Al–Mn alloy observed by high resolution microscopy. *Sci. Rep. Res. Inst. Tohoku Univ. A* 32 (1985) 309–14.

Hiraga, K., Sugiyama, K. & Ohsuna, T. Atom cluster arrangements in cubic approximant phase of icosahedral quasicrystals. *Phil. Mag. A* 78 (1998) 1051–64.

Hiraga, K., Ohsuna, T. & Sugiyama, K. Atom clusters with icosahedral symmetry in cubic alloy phases related to icosahedral quasicrystals. *The Rigaku J.* 16 (1999) 38–45.

Hoare, M. R. & Pal, P. Physical cluster mechanics: statics and energy surfaces for monatomic systems. *Adv. Phys.* 20 (1971) 161–98.

Hobbs, L. W., Jesurum, C. E., Pulim, V. & Berger, B. Local topology of silica networks. *Phil. Mag. A* 78 (1998) 679–711.

Hoffman, D. The computer-aided discovery of new embedded minimal surfaces. *Math. Intelligencer.* 9 (1987) 8–21.

Hooke, R. *Micrographia: Or, Some Physiological Descriptions of Minute Bodies Made by Magnifying Glasses* (Martyn, J. & Allestry, J., eds). Martyn & Allestry, London (1665).

Hren J. J. & Ranganathan S. (eds) *Field-Ion Microscopy*. Plenum Press, New York (1967), Russian edition Mir Publishers, Moscow (1971).

Hsiang, W. Y. & Hsiang, W.-Y. *Least Action Principle of Crystal Formation of Dense Packing Type and the Proof of Kepler's Conjecture*. World Scientific, Singapore (2002).

Hudson, D. R. Density and packing in an aggregate of mixed spheres. *J. App. Phys.* 20 (1949) 154–62.

Hume-Rothery, W. Researches on the nature, properties, and condition of formation of intermetallic compounds. *J. Inst. Metals* 35 (1926) 319–335.

Huntley, H. E. *The Divine Proportion*. Dover, New York (1970).

Hyde, S. T. The topology and geometry of infinite periodic surfaces. *Z. Krystallogr.* 187 (1989) 165–85.

Hyde, S. T. Hyperbolic surfaces in the solid state and the structure of ZSM-5 zeolites. *Acta Chemica Scand.* 45 (1991) 860–63.

Hyde, S. T. & Andersson, S. Differential geometry of crystal structure description: relationships and phase transitions. *Z. Kristallogr.* 170 (1985) 225–39.

Hyde, S. T. & Ramsden, S. Crystals: two-dimensional non-Euclidean geometry and topology. Chapter 2 of *Mathematical Chemistry*, vol. 6 (Bonchev, D. & Rouvray D, eds). Gordon and Breach, New York (2000a).

Hyde, S. T. & Ramsden, S. Polycontinuous morphologies and interwoven helical networks. *Europhys. Lett.* 50 (2000b) 135–41.

Hyde, S. T. & Ramsden, S. Some novel three-dimensional Euclidean crystalline networks derived from two-dimensional hyperbolic tilings. *European Physical Journal E* 31 (2003) 273–84.

Hyde, S. T., Andersson, S., Ericsson, B. & Larsson, K. A systematic net description of saddle polyhedra and periodic minimal surfaces of the gyroid type in the glycerolmonooleate-water system. *Z. Kristallogr.* 168 (1984) 221–54.

Hyde, S. T., Andersson, S., Larsson, R., Blum, Z., Landh, T., Lidin, S. & Ninham, B. W. *The Language of Shape: The Role of Curvature in Condensed Matter: Physics, Chemistry and Biology*. Elsevier (1996).

Iijima, S. Helical microtubules of graphitic carbon. *Nature* 354 (1991) 56–8.

Inoue, A., High strength bulk amorphous alloys with low critical cooling rate. *Mater. Trans. JIM* 36, (1995) 866–75.

Inoue, A *Bulk Amorphous Alloys – Preparation and Fundamental Characteristics*. Trans Tech Publications, Zürich (1998).

Inoue, A. Stabilization of metallic supercooled liquid and bulk amorphous alloys. *Acta Mater.* 48 (2000) 279–306.

Janot, Ch. & Patera, J. Simple physical generation of aperiodic structures. *J. Non-Cryst. Solids* 233–4 (1998) 234–8.

Jeanneret, C. E. ("le Corbusier"). *The Modulor: A Harmonious Measure to the Human Scale Universally Applicable to Architecture and Mechanics and Modulor 2 (Let the User Speak Next).* Birkhäuser (2000) [Facsimile of 1st English Edition, Faber and Faber (1954)].

Jeong, H.-C. Inflation rule for Gummelt coverings with decorated decagons and its implication to quasi-unit-cell models. *Acta Cryst. A* 59 (2003) 361–6.

Jeong, H.-C. & Steinhardt, P. J. Constructing Penrose-like tilings from a single prototile and the implications for quasicrystals. *Phys. Rev. B* 55(1997) 3520–32.

Johnson, W. L. Bulk glass forming metallic alloys: science and technology. *Mat. Res. Bull.* 24 (1999) 42–56.

Johnson, R. L. *Atomic and Molecular Clusters.* Taylor & Francis, London and New York (2002).

Johnson, C. K., Burnett, M. N. & Dunbar, W. D. Crystallographic topology and its applications. In: *Crystallographic Computing 7: Macromolecular Crystallographic Data* (Bourne, P. E. & Watenpaugh, K. D., eds). Oxford University Press (2002).

Kapraff, J. The relationship between mathematics and mysticism of the golden mean through history. In: *Fivefold Symmetry* (Hargittai, I., ed.). World Scientific, Singapore (1992).

Karcher, H. The triply periodic minimal surfaces of Alan Schoen and their constant mean curvature companions. *Manuscripta Mathematica* 64 (1989a) 291–357.

Karcher, H. *Construction of Minimal Surfaces. Vorlesungsreihe,* vol. 12. Institut für Angewandte Mathematik, Univ. Bonn (1989b).

Karcher, H. & Polthier, K. Construction of triply periodic minimal surfaces. *Phil. Trans. Roy. Soc. London A* 354 (1996) 1–18.

Katz, A. Theory of matching rules for the 3-dimensional Penrose tilings. *Commun. Math. Phys.* 118 (1986) 263–88.

Kelvin, W. T. & Weaire, D. *The Kelvin Problem: Foam Structures of Minimal Surface Area.* Taylor & Francis, London (1997).

Kepler, J. *Mysterium Cosmographicum.* Tübingen (1596, 1621).

Kepler, J. *Strena seu de Nive Sexangula,* Tampach, Frankfort (1611); *The Six-cornered Snowflake.* Clarendon Press, Oxford (1966).

Kepler, J. *Harmonice Mundi.* Lincii Austriae, Sumptibus Godfrdi Tampachhii, Francof. (1619); German transl. *Johannes Kepler, Gesammelte Werke* (Caspar, M. ed.). Beck, Munich (1990). *The Harmony of the World.* English translation with Introduction and Notes by Alton, E. J., Duncan, A. M. & Field, J. V. *Memoirs of the American Philosophical Society* 209 (1997).

Kikuchi, R. Shape distribution of two-dimensional soap froths. *J. Chem. Phys.* 24 (1956) 862–7.

Klinowski, J., Mackay, A. L. & Terrones, H. Curved surfaces in chemical structures. *Phil. Trans. Roy. Soc. London A* 354 (1996) 1975–87.

Koch, E. Minimal surfaces with self-intersections along straight lines. II. Surfaces forming three-periodic labyrinths. *Acta Cryst. A* 56 (2000) 15–23.

Koch, E. & Fischer, W. Zur Bestimmung asymmetrische Einheiten kubischer Raumgruppen mit Hilfe von Wirkungsbereiche. *Acta Cryst. A* 30 (1974) 490–6.

Koch, E. & Fischer, W. On 3-periodic minimal surfaces with non-cubic symmetry. *Z. Kristallogr.* 183 (1988) 129–52.

Koch, E. & Fischer, W. New surface patches for minimal balance surfaces. II. Multiple catenoids. *Acta Cryst. A* 45 (1989a) 169–74; IV. Catenoids with spout-like attachments. *Acta Cryst. A* 45 (1989b) 558–63.

Koch, E. & Fischer, W. Flat points for minimal balance surfaces. *Acta Cryst. A* 46 (1990) 33–40.

Koch, E. & Fischer, W. Triply periodic minimal balance surfaces: a correction. *Acta Cryst. A* 49 (1993) 209–10.

Koch, E. & Fischer, W. Sphere packings with three contacts per sphere and the problem of the least dense packing. *Z. Kristallogr.* 210 (1995) 407–14.

Koch, E. & Fischer, W. Minimal surfaces with self-intersections along straight lines. I. Derivation and properties. *Acta Cryst. A* 55 (1999) 58–64.

Komura, Y., Sly, W. G. & Shoemaker, D. P. The crystal structure of the R phase, Mo-Cu-Cr. *Acta Cryst.* 13 (1960) 575–85.

Kottwitz, D. A. The densest packing of equal circles on a sphere, *Acta Cryst. A* 47 (1991) 158–65.

Kovacs, F. & Tarnai, T. An expandable dodecahedron. *Hyperspace* 10(1) (2001) 13–20.

Kowalewski, G. *Der Keplersche Körper und andere Bauspiele.* Koehlers, Leipzig (1938). English translation: *Construction Games with Kepler's Solids,* tr. D. Booth. Parker Courtney Press, Chestnut Ridge, NY (2001).

Kramer, P. Non-periodic central space-filling with icosahedral symmetry using copies of seven elementary cells. *Acta Cryst.* A38 (1982) 257–64.

Kramer, P. & Neri, R. On periodic and non-periodic space fillings of E^m by projection. *Acta Cryst. A* 40 (1984) 580–7.

Kreiner, G. & Franzen, H. F. A new cluster concept and its application to quasi-crystals of the i-AlMnSi family and closely related crystalline structures. *J. Alloy Compd.* 221 (1995) 15–36.

Kreiner, G. & Schäpers, M. A new description of Samson's Cd_3Cu_4 and a model of icosahedral i-CdCu. *J. Alloys Compd.* 259 (1997) 83–114.

Kripyakevich, P. I. A systematic classification of types of intermetallic structures. *J. Struct. Chem.* (1963) 1–35.

Kripyakevitsch, P. I. *Structurtypen der Intermetallischen Phasen.* Verlag Nauka, Moskau (1977).

Kroto, H. W., Heath, J. R., Obrien, S. C., Curl, R. F. & Smalley, R. E. C-60 – Buckminsterfullerene. *Nature* 318 (1985) 162–3.

Krypyakevich, P. I. A systematic classification of types of intermetallic structures. *J. Struct. Chem.* (1963) 1–35.

Kuijlaars, A. B. J. & Saff, E. B. Asymptotics for minimal discrete energy on the sphere. *Tran. Amer. Math. Soc.* 350 (1998) 523–38.

Kuo, K. H. Mackay, anti-Mackay, double-Mackay, pseudo-Mackay, and related icosahedral shell clusters. *Structural Chemistry* 13 (2002) 221–9.

Lambert, C. A., Radzilowski, L. H. & Thomas, E. L. Triply periodic level surfaces as models for cubic tricontinuous block copolymer morphologies. *Phil. Trans. Roy. Soc. London A* 354 (1996) 2009–23.

Leech, J. The problem of the thirteen spheres. *Math. Gazette* 40 (1956) 22–3.

Lehninger, L., Nelson, D. L. & Cox, M. M. *Principles of Biochemistry.* Worth publishers, New York (1993).

Leoni, S. & Nesper, R. Elucidation of simple pathways for reconstructive phase transitions using periodic equi-surface (PES) descriptors. The silica phase system. I. Quartz-tridymite. *Acta Cryst. A* 56 (2000) 383–93.

Levine, D. & Steinhardt, P. J. Quasicrystals: a new class of ordered structures. *Phys, Rev. Lett.* 53 (1984) 2477–80.

Li, C. R., Ranganathan S. & Inoue, A. Initial crystallization processes of Hf–Cu–M (M = Pd, Pt or Ag) amorphous alloys. *Acta Mater.* 49 (2001) 1903–8.

Li, C. R., Zhang, X. N. & Cao, Z. X. Triangular and Fibonacci number patterns driven by stress on core/shell microstructures. *Science* 309 (2005) 909–11.

Li, H. Eddaoudi, M., O'Keefe, M. & Yaghi, O. M. Design and synthesis of an exceptionally stable and highly porous metal-organic framework. *Nature* 402 (1999) 276–9.

Li, X. Z. Structure of Al–Mn decagonal quasicrystal. I. A unit-cell approach. *Acta Cryst. B* 51 (1995) 265–70.

Li, X. Z. & Frey, F. Structure of Al–Mn decagonal quasicrystal. II. A high-dimensional description. *Acta Cryst. B* 51 (1995) 271–5.

Lidin, S. Ring-like minimal surfaces. *J. Phys. France* 49 (1988) 421–7.

Lidin, S. & Andersson, S. Regular polyhedral helices. *Z. anorg. Alg. Chem.* 622 (1996) 164–6.

Lidin, S. & Larsson, S. Bonnet transformation of infinite periodic minimal surfaces with hexagonal symmetry. *J. Chem. Soc. Faraday Trans.* 86 (1990) 769–75.

Lidin, S., Hyde, S. T. & Ninham, B. W. Exact construction of periodic minimal surfaces: the I-WP surface and its isometries. *J. Phys. France* 51 (1990) 801–13.

Liebau, F., Gies, H., Gunawardena, R. P. & Marier, B. Classification of tectosilicates and systematic nomenclature of clathrate type tectosilicates: a proposal. *Zeolites* 6 (1986) 373–7.

Lifshitz, R. Theory of color symmetry for periodic and quasiperiodic crystals. *Physica A* 232 (1996) 633–47.

Lifshitz, R. Lattice color groups of quasicrystals. *Phys. Rev. Lett.* 80 (1998) 2717–20.

Lindenmayer, A. Mathemathical models for cellular interaction in development. *J. Theor. Biol.* 18 (1968) 280–315.

Livio, M. *The Golden Ratio: The Story of PHI, the World's Most Astonishing Number.* Broadway Books, New York (2003).

Loeb, A. L. A modular algebra for the description of crystal structures. *Acta Cryst.* 15 (1962) 219–26.

Loeb, A. L. Moduledra crystal models. *Amer. J. Phys.* 31 (1963) 190–6.

Loeb, A. L. A systematic survey of cubic crystal structures. *J. Solid State Chem.* 1 (1970) 237–67.

Loeb, A. L. *Space Structures – their Harmony and Counterpoint.* Addison-Wesley, New York (1974).

Loeb, A. L. Hierarchical structure and pattern recognition in minerals and alloys. *Per. Mineral.* 59 (1990) 197–217.

Longuet-Higgins, M. S. *Learning and Exploring with RHOMBO.* Dextro Mathematical toys, Del Mar, USA (1992).

Longuet-Higgins, M. S. Nested triacontahedral shells or how to grow a quasicrystal. *Math. Intell.* 25 (2003) 25–43.

Lord, E. A. Quasicrystals and Penrose patterns. *Curr. Sci.* 61 (1991) 313–9.

Lord, E. A. Triply periodic balance surfaces. *Colloids and Surfaces A* 129–30 (1997) 279–95.

Lord, E. A. Helical structures: the geometry of protein helices and nanotubes. *Structural Chemistry* 13 (2002) 305–14.

Lord, E. A. & Mackay, A. L. Periodic minimal surfaces of cubic symmetry. *Curr. Sci.* 85 (2003) 346–62.

Lord, E. A. & Ranganathan, S. The Gummelt decagon as a quasi unit cell. *Acta Cryst. A* 57 (2001a) 531–9.

Lord, E. A. & Ranganathan, S. Sphere packing, helices and the polytope $\{3, 3, 5\}$. *EPJ D* 15 (2001b) 335–43.

Lord, E. A. & Ranganathan, S. The γ-brass cluster and the Boerdijk–Coxeter helix. *J. Non-Cryst. Solids* 334–5 (2004) 121–5.

Lord, E. A. & Wilson, C. B. *The Mathematical Description of Shape and Form.* Ellis Horwood, Chichester (1984).

Lord, E. A., Ranganathan, S. & Kulkarni, U. D. Tilings, coverings, clusters and quasicrystals. *Curr. Sci.* 78 (2000) 64–72.

Lord, E. A., Ranganathan, S. & Kulkarni, U. D. Quasicrystals: tiling versus clustering. *Phil. Mag. A* 81 (2001) 2645–51.

Louzguine, D., Ko, M. S., Ranganathan, S. & Inoue, A. Nano-crystallization of the Fd$\bar{3}$d Ti$_2$Ni type phase in Hf based metallic glasses. *J. Nasnosci. Nanotech.* 1 (2001) 185–90.

Lück, R. Dürer–Kepler–Penrose: the development of pentagonal tilings. *Mater. Sci. Eng.* 82 (2000) 263–7.

Lucretius (Titus Lucretius Carus). *De Rerum Natura.* (English translation by Latham, R. E., ed.). Penguin, Harmondsworth, Middlesex, UK (1994).

Mackay, A. L. A dense non-crystalline packing of equal spheres. *Acta Cryst.* 15 (1962) 1916–8.

Mackay, A. L. Generalised crystallography. *Izvj. Jugosl. Centr. Krist. (Zagreb)* 10 (1975) 15–36.

Mackay, A. L. Crystal symmetry. *Phys. Bull.* November (1976) 495–7.

Mackay, A. L. De nive quinquangula: on the pentagonal snowflake. *Sov. Phys. Cryst.* 26 (1981) 517–22.

Mackay, A. L. Crystallography and the Penrose pattern. *Physica* 114A (1982) 609–13.

Mackay, A. L. What has Penrose tiling to do with the icosahedral phases? Geometrical aspects of the icosahedral quasicrystal problem. *J. Microsc.* 146 (1987) 233–43.

Mackay, A. L. New geometries for superconduction and other purposes. *Speculat. Sci. Technol.* 11 (1988) 4–8.

Mackay, A. L. Geometry of interfaces. *Colloque de Physique.* 51 C7 (1990) 399–405.

Mackay, A. L. Crystallographic surfaces. *Proc. Roy. Soc. London A* 442 (1993) 47–59.

Mackay, A. L. Periodic minimal surfaces from finite element methods. *Chem. Phys. Lett.* 221 (1994) 317–21.

Mackay, A. L. Flexicrystallography: curved surfaces in chemical structures. *Curr. Sci.* 69 (1995) 151–60.

Mackay, A. L. Tools for thought. *Revista Especializada en Ciencias Quimico-Biologicas* 2 (1999) 69–74.

Mackay, A. L. The shape of two-dimensional space. Chapter 14 of *Symmetry 2000* (Hargittai, I., ed.). Portland Press, London (2000).

Mackay, A. L. Generalised crystallography. *Structural Chemistry* 13 (2002) 215–20.

Mackay, A. L. & Terrones, H. Diamond from graphite. *Nature* 352 (1991) 762.

Mackay, A. L. & Terrones, H. Hyperbolic graphitic structures with negative Gaussian curvature. *Phil. Trans. Roy. Soc. London A* 343 (1993) 113–27.

Mackay, A. L. Finney, J. L. & Gotoh, K. The closest packing of equal spheres on a spherical surface. *Acta Cryst. A* 33 (1977) 98–100.

Mackintosh, A. R. The Fermi surface of metals. *Sci. Am.* 209 (1963) 110–20.

Mandelbrot, B. *The Fractal Geometry of Nature.* W. H. Freeman & Co. New York (1982).

Manoharan, V. N., Elsesser, M. T. & Pine, D. J. Dense packing and symmetry in small clusters of microspheres. *Science* 301 (2003) 483–7.

Margulis, L., Salitra, G., Tenne, R. & Talianker, M. Nested fullerene-like structures. *Nature* 365 (1993) 113–4.

Marzec, C. & Kapraff, J. Properties of maximal spacing on a circle relating to phyllotaxis and the golden mean. *J. Theor. Biol.* 103 (1983) 201–26.

Měch, R. & Prusinkiewicz, P. Visual models of plants interacting with their environment. *Proceeding of SIGGRAPH '96 (New Orleans, August, 4–9 1996).* Computer Graphics Proceedings, Annual Conference Series, ACM SIGGRAPH (1996) 397–410.

Meier, W. M. & Moeck, H. J. The topology of three-dimensional 4-connected nets; classification of zeolite framework types in terms of coordination sequences. *J. Non-Cryst. Solids* 27 (1979) 349–55.

Melnyk, T. W., Knop, O. & Smith, W. R. Extremal arrangements of points and unit charges on a sphere. *Can. J. Chem.* 53 (1977) 1745–61.

Mikalkovic, M., Zhu, W.-J., Henley, C. L. & Oxborrow, M. Icosahedral quasicrystal decoration models: I. Geometrical principles. *Phys. Rev. B* (1996) 9002–20.

Miracle, D. B. A structural model for metallic glasses. *Nature Mater.* 3 (2004) 697–702.

Miracle, D. B., Sanders, W. S. & Senkov, O. N. The influence of efficient atomic packing on the constitution of metallic glasses. *Phil. Mag.* 83 (2003) 2409–28.

Miyazaki, K. *On Some Periodical and Non-periodical Honeycombs.* University of Kobe, Japan (1977a).

Miyazaki, K. *Polyhedra and Architectures* (in Japanese). Shokokusha publishers, Tokyo (1977b).

Miyazaki, K. *Form in space: Polygons, Polyhedra and Polytopes.* Asaku Publication Company, Tokyo (1983).

Miyazaki, K. *An Adventure in Multidimensional Space: The Art and Geometry of Polygons, Polyhedra and Polytopes.* Wiley, New York (1986).

Miyazaki, K. & Takada, I. Uniform ant-hills in the world of golden zonohedra. *Structural Topology* 4 (1980) 21–9.

Miyazaki, K. & Yamagiwa, T. *On the Golden Polytopes.* University Kobe, Japan (1977).

Molnár, E. On triply periodic minimal balance surfaces. *Structural Chemistry* 13 (2002) 267–76.

Moore, A. J. W. & Ranganathan, S. The interpretation of field ion images. *Phil. Mag.* 16 (1967) 723–37.

Moore, P. G. & Smith, J. V. Archimedean polyhedra as the basis of tetrahedrally coordinated frameworks. *Mineralogic. Mag.* 33 (1964) 1008–14.

Mueller, E. W. & Tsong, T. T. *Field Ion Microscopy.* American Elsevier Publishing Company, New York (1969).

Nath, M. & Rao, C. N. R. New metallic disulphide nanotubes. *J. Amer. Chem. Soc.* 123 (2001) 4841–2.

Naylor, M. Golden, $\sqrt{2}$, and π flowers: a spiral story. *Mathematics Magazine* 75 (2002) 163–72.

Nelson, D. R. & Spaepen, F. *Polytetrahedral Order in Condensed Matter.* Academic Press, New York (1989).

Neovius, E. R. Bestimmung zweier speciellen periodischen Minimalflächen. *Helsingfors Akad. Abhandlungen* (1883).

Nesper, R. Bonding patterns in intermetallic compounds. *Angew. Chem. Int. Ed. Engl.* 30 (1991) 789–817.

Nesper, R. & Leoni, S. On tilings and patterns on hyperbolic surfaces and their relation to structural chemistry. *Chemphyschem* 2 (2001) 413–422.

Nitsche, J. C. C. *Lectures on Minimal Surfaces*, vol. 1. Cambridge University Press (1989).

Nyman, H. & Andersson, S. On the structure of Mn_5Si_3, Th_6Mn_{23} and γ-brass. *Acta Cryst.* A 35 (1979) 580–3.

Nyman, H. & Hyde, B. G. The related structures of α-Mn, sodalite, Sb_2Tl_7, etc. *Acta Cryst.* A 37 (1981) 11–17.

Nyman, H., Andersson, S., Hyde, B. G. & O'Keeffe, M. The pyrochlore structure and its relatives. *J. Solid State Chem.* 26 (1978) 123–31.

Nyman, H., Carroll, C. E. & Hyde, B. G. Rectilinear rods of face-sharing tetrahedra and the structure of β-Mn. *Z. Kristallogr.* 196 (1991) 39–46.

Oberteuffer, J. A. & Ibers, J. A. A refinement of the atomic and thermal parameters of α-manganese from a single crystal. *Acta Cryst.* B26 (1970) 1499–504.

Ogawa, T. In: *Proceedings of the First International Symposium for Science on Form.* KTK Sc. Publishers, Tokyo (1986) 479.

Oguey, C. & Sadoc, J.-F. Crystallographic aspects of the Bonnet transformation for periodic minimal surfaces (and crystals of films). *J. Phys. I France* 51 (1993) 839–54.

Okabe, A., Boots, B., Sugihara, K. & Sung, N. C. *Spatial Tessellations: Concepts and Applications of Voronoi Diagrams.* Wiley, New York (2000).

O'Keeffe, M. Cubic cylinder packings. *Acta Cryst. A* 48 (1992) 879–84.

O'Keeffe, M. Uninodal 4-connected 3D nets. II. Nets with 3-rings. *Acta Cryst. A* 48 (1992) 670–3.

O'Keeffe, M. Uninodal 4-connected 3D nets. III. Nets with three or four 4-rings at a vertex. *Acta Cryst. A* 51 (1995) 916–20.

O'Keeffe, M. New ice outdoes related nets in smallest ring size. *Nature* 392 (1998a) 879.

O'Keeffe, M. Sphere packings and space filling by congruent simple polyhedra. *Acta Cryst. A* 54 (1998b) 320–9.

O'Keeffe, M. 4-Connected nets of packings of non-convex parallelohedra and related simple polyhedra. *Z. Kristallogr.* 214 (1999) 438–42.

O'Keeffe, M. & Brese, N. E. Uninodal 4-connected 3D nets. I. Nets without 3- or 4-rings. *Acta Cryst. A* 48 (1992) 663–9.

O'Keeffe, M. & Hyde, B. G. Vertex symbols for zeolite nets. *Zeolites* 19 (1997) 370–4.

O'Keeffe, M. & Hyde, B. G. Plane nets in crystal chemistry. *Proc. Roy. Soc. London A* 295 (1980) 553–618.

O'Keeffe, M. & Hyde, S. T. Vertex symbols for zeolite nets. *Zeolites* 19 (1997) 370–4.

O'Keeffe, M., Eddaoudi, M., Li, H., Reineke, T. & Yaghi, O. M. Frameworks for extended solids: geometrical design principles. *J. Solid state Chem.* 152 (2000) 3–20.

O'Keeffe, M., Plévert, J., Teshima, Y., Watanabe, Y. & Ogama, T. The invariant cubic rod (cylinder) packings: symmetries and coordinates. *Acta Cryst. A* 57 (2001) 110–11.

O'Keeffe, M., Plévert, J. & Ogawa, T. Homogeneous cubic cylinder packings revisited. *Acta Cryst. A* 58 (2002) 125–32.

Osserman, R. *A Survey of Minimal Surfaces.* Dover, New York (1986).

Ozin, G. A. Curves in chemistry: supramolecular materials taking shape. *Can. J. Chem.* 77 (1999) 2001–14.

Pacioli, L. *De Divina Proportione.* Venice (1509); Abaris Books, New York (2004).

Pauling, L. The crystal structure of magnesium stannide. *J. Am. Chem. Soc.* 45 (1923) 2777–80.

Pauling, L. The stochastic method and the structure of proteins. *Am. Sci.* 43 (1955) 285–97.

Pauling, L. *The Nature of the Chemical Bond and the Structure of Molecules and Crystals: An Introduction to Modern Structural Chemistry* (3rd ed). Cornell University Press, Ithaca, New York (1960).

Pauling, L. The close-packed spheron model of nuclear fission. *Science* 150 (1965) 297–305.

Pauling L. & Marsh, R. E. The structure of chlorine hydrate. *Proc. Nat. Acad. Sci. Wash.* 38 (1952) 112–18.

Peano, G. Sur une courbe qui remplit toute une aire plane. *Math. Annalen* 36 (1890) 157–60.

Pearce, P. *Structure in Nature is a Strategy for Design.* MIT Press, Cambridge, MA (1978).

Peitgen, H.-O. & Richter, P. *The Beauty of Fractals.* Springer, New York (1986).

Penrose, R. The role of aesthetics in pure and applied mathematical research. *Bull. Inst. Math. Appl.* 10 (1974) 266–71.

Penrose, R. Pentaplexity: a class of non-periodic tilings of the plane. *Eureka* 39 (1978) 16–22; reprinted in *Math. Intell.* 2 (1979) 32–8.

Pettifor, D. A chemical scale for crystal-structure maps. *Solid State Commun.* 51 (1984) 37–4.

Pettigrew, J. B. *Design in Nature.* Longmans, London (1908).

Phillips, E. G. *An Introduction to Crystallography.* Longmans, London (1946; 2nd edn 1956).

Pisano, L. (Fibonacci). *Liber Abaci* (1202); first English translation (Sigler, L. E., ed). Springer, New York (2002).

Plateau, J. A. F. *Statique Expérimentale et Théorique des Liquides Soumis aux Seules Forces Moléculaires.* Gauthier-Villars, Paris (1873).

Poinsot, L. Sur les polygones et les polyèdres. *J. École Polytech.* 10 (1810) 16–48.

Prusinkiewicz, P. & Lindenmayer, A. *The Algorithmic Beauty of Plants.* Springer, New York (1990).

Rakhmanov, E. A., Saff, E. B. & Zhou, Y. M. Minimal discrete energy on the sphere. *Math. Res. Lett.* 1 (1994) 647–62.

Rakhmanov, E. A., Saff, E. B. & Zhou, Y. M. Electrons on the sphere. In: *Computational Method and Function Theory* (Ali, R. M., Rascheweyh, St. & Saff, E. B., eds). World Scientific, Singapore (1995).

Ramachandra Rao, P. & Sastry, G. V. S. A basis for synthesis of quasicrystals. *Pramana* 25 (1985) L225–30.

Ranganathan, S. & Chattopadhyay, K. Quasicrystals. *Ann. Rev. Mater. Sci.* 21 (1991) 437–62.

Ranganathan, S., Chattopadhyay, K., Singh, A. & Kelton, K. F. Decagonal quasicrystals. *Prog. Mater. Sci.* 41 (1997) 195–240.

Ranganathan, S., Divakar, R. & Raghunathan, V. S. Interface structures in nanocrystalline materials. *Scr. Mater.* 44 (2001) 1169–74.

Ranganathan, S., Singh, A. & Tsai, A. P. Frank's 'cubic' hexagonal phase: an intermetallic cluster compound as an example. *Phil. Mag. Lett.* 82 (2002) 13–19.

Rao, C. N. R. & Gopalakrishnan, J. *New Directions in Solid State Chemistry.* Cambridge University Press (1997).

Rivier, N. Kelvin's conjecture on minimal froths and the counter-example of Weaire and Phelan. *Forma* 11 (1996) 195–8.

Robinson, R. M. Arrangement of 24 points on a sphere. *Math. Ann.* 144 (1961) 17–48.

Robinson, R. M. Finite sets of points on a sphere with each nearest to five others. *Math. Ann.* 179 (1969) 296–318.

Rogers, C. A. The packing of equal spheres. *Proc. London Math. Soc.* 8 (1958) 609–20.

Romeu, D. & Aragon, J. L. Quasicrystals and their approximants. In: *Crystal-Quasicrystal Transitions* (Yacamán, M. J. & Torres, M., eds). Elsevier, New York (1993) 193–215.

Runnels, L. K. Ice. *Sci. Am.* 215 (1966) 118–26.

Sadoc, J. F. Helices and helix packings derived from the {3, 3, 5} polytope. *Eur. Phys. J. E* 5 (2001) 575–82.

Sadoc, J. F. & Charvolin, J. Infinite periodic minimal surfaces and their crystallography in the hyperbolic plane. *Acta Cryst. A* 45 (1989) 10–20.

Sadoc, J. F. & Mosseri, R. The E8 lattice and quasicrystals: geometry, number theory and quasicrystals. *J. Phys. A: Math Gen.* 26 (1993) 1789–809.

Sadoc, J. F. & Mosseri, R. *Geometrical Frustration.* Cambridge University Press (1999).

Sadoc, J. F. & Rivier, N. (eds) *Foams and Emulsion.* Kluwer Academic Press, Dordrecht, the Netherlands (1999).

Sadoc, J. F. & Rivier, N. Boerdijk–Coxeter helix and biological helices. *Eur. Phys. J. B* 12 (1999) 309–18.

Saff, E. B. & Kuijlaars, A. B. J. Distributing many points on a sphere. *Math. Intell.* 19 (1997) 5–11.

Samson, S. The crystal structure of the phase β of Mg_2Al_3. *Acta Cryst.* 19 (1965) 401–13.

Samson, S. The crystal structure of the intermetallic compound Cu_4Cd_3. *Acta Cryst.* 23 (1967a) 586–600.

Samson, S. The structure of complex intermetallic compounds. In: *Structural Chemistry and Molecular Biology* (Rich, A. & Davidson, N., eds). Freeman, New York (1968) 687–717.

Samson, S. Structural principles of giant cells. In: *Developments in the Structural Chemistry of Alloy Phases* (Giessen, B. C., ed.). Plenum Press, New York (1969) 65–106.

Samson, S. Complex cubic A_6B compounds. II. The crystal structure of Mg_6Pd. *Acta Cryst.* B 28 (1972) 936–45.

Sander, L. M. Fractal growth. *Sci. Am.* 256 (1987) 94–100.

Schattschneider, D. The plane symmetry groups, their recognition and notation. *Amer. Math. Month.* 85 (1978) 439–58.

Schattschneider, D. In black and white: how to create perfectly colored symmetric patterns. *Comp. Maths with Appl.* 12B (1986) 873–95.

Scheffer, M. & Lück, R. Coloured quasiperiodic patterns with tenfold symmetry and eleven colours. *J. Non-Cryst. solids* 250–2 (1999) 815–9.

Scherk, H. F. Bemerkung über der kleinste Fläche innerhalb gegebener Grenzen. *J. reine angew. Math.* 13 (1834) 185–208.

Schindler, M., Hawthorne, F. C. & Baer, W. H. Metastructure: homeomorphisms between complex inorganic structures and three-dimensional nets, *Acta Cryst.* B 55 (1999) 811–29.

Schoen, A. H. Homogeneous nets and their fundamental regions. *Not. Amer. Math. Soc.* 14 (1967) 661.

Schoen, A. H. Regular saddle polyhedra. *Not. Amer. Math. Soc.* 15 (1968) 929.

Schoen, A. H. Infinite periodic minimal surfaces without self-intersections, *NASA Technical Note TN D-5541* (1970) i–vii; 1–98.

Schrandt, R. G. & Ulam, S. In: *Essays on cellular Automata* (Burks, A. W., ed.). University of Illinois Press (1970).

Schrödinger, E. *What is Life?* Cambridge University Press (1944); 2nd edn (1992).

Schwarz, H. A. *Gesammelte Mathematische Abhandlungen I.* Springer, Berlin (1890).

Schwarz, U. S. & Gompper, G. Systematic approach to bicontinuous cubic phases in ternary amphiphilic systems. *Phys. Rev. E* 59 (1999) 5528–41.

Scott, G. D. & Kilgour, D. M. The density of random close packing of spheres. *J. Phys. D: Appl. Phys.* 2 (1969) 863–6.

Seifert, G., Terrones, H., Terrones, M. & Frauenheim, T. Novel NbS_2 metallic nanotubes. *Solid State Commun.* 115 (2000) 635–8.

Senechal, M. Geometry and crystal symmetry. *Comp. Maths with Appl.* 12B (1986) 565–78.

Senechal, M. & Fleck, G. (eds). *Shaping Space: a Polyhedral Approach.* Springer, New York (1988).

Senkov, O. N. & Miracle, D. B. Effect of the atomic size distribution on glass forming ability of amorphous metallic alloys. *Mater. Res. Bull.* 36 (2001) 2183–98.

Shechtman, D., Blech, I., Gratias D. & Cahn, J. W. Metallic phase with long-ranged orientational order and no translational symmetry. *Phys Rev. Lett.* 53 (1984) 1951–3.

Sheng, H. W., Luo, W. K., Alamgir, F. M., Bai, J. M. & Ma, E. Atomic packing and short-to-medium range order in metallic glasses. *Nature* 439 (2006) 419–25.

Shipman, P. D. & Newell, A. C. Phyllotactic patterns in plants. *Phys. Rev. Lett.* 92, 168102 (2004) 1–4.

Shubnikov, A. V., Belov, N. V. *et al. Coloured Symmetry.* Pergamon Press, Oxford (1963).

Shukla, K.S. *The Aryabhatiya of Aryabhata with the Commentary of Bhaskara I and Somesvara.* Indian National Science Academy, New Delhi (1976).

Skilling, J. The complete set of uniform polyhedra. *Phil. Trans. Roy. Soc. London A* 278 (1975) 111–35.

Sloane, N. J. A. Kepler's conjecture confirmed. *Nature* 395 (1998) 435–6.

Sloane, N. J. A., Hardin, R. H., Duff, T. D. S. & Conway, J. H. Minimal energy clusters of hard spheres. *Discrete and Computational Geometry* 14 (1995) 237–59.

Smalley, R. E. & Curl, R. F. The fullerenes. *Sci. Am.* 265 (1991) 32–7.

Smith, C. S. Further notes on the shape of metallic grains: space-filling polyhedra with unlimited sharing of corners and faces. *Acta Metallurgica* 1 (1953) 295–300.

Smith, C. S. The tiling patterns of Sebastian Truchet and the topology of structural hierarchy. *Leonardo* 20 (1987) 373–85.

Steinhardt, P. J. & Ostlund, S. (eds). *The Physics of Quasicrystals.* World Scientific, Singapore (1987).

Steinhardt, P. J., Nelson, D. R. & Ronchetti, M. Bond-orientational order in liquids and glasses. *Phys. Rev. B* 28 (1983) 784–805.

Steinhardt, P. J., Jeong, H.-C., Saitoh, K., Tanaka, M., Abe, E. & Tsai, A. P. Experimental verification of the quasi-unit-cell model of quasicrystal structure. *Nature* 396 (1998) 55–7.

Stessmann, B. Periodische Minimalflächen. *Math. Zeitschrift* 38 (1934) 417–42.

Steurer, W., Haibach, T. & Zhang, B. The Structure of decagonal Al_{70} Ni_{15} Co_{15}. *Acta Cryst.* B49 (1993) 661–75.

Steurer, W. The structure of quasicrystals. In: *Physical Metallurgy* (Cahn, R. W. & Haasen, P., eds). North-Holland, Amsterdam (1996).

Stevens, P. S. *Patterns in Nature.* Penguin, Harmondsworth, Middlesex, UK (1974); Little, Brown, New York (1979).

Stewart, I. Crystallography of a golf ball. *Sci. Am.* February (1997) 80–2.

Stiny, G. *Pictorial and Formal Aspects of Shape and Shape Grammars.* Birkhäuser, Basel (1975).

Stone, A. J. & Wales, D. J. Theoretical studies of icosahedral C60 and some related species. *Chem. Phys. Lett.* 128 (1986) 501–3.

Sugiyama, K., Yasuda, K., Ohsuna, T. & Hiraga, K. The structure of the hexagonal phases in Mg–Zn–RE (RE = Sm and Gd) alloys. *Z. Kristallogr.* 213 (1998) 537–43.

Sullinger, D. B. & Kennard, C. H. L. Boron Crystals. *Sci. Am.* July (1966) 96–107.

Sutton, D. *Platonic and Archimedean Solids.* Walker & Co., New York (2002).

Swinton, J. Watching the daisies grow: Turing and Fibonacci Phyllotaxis. In: *Alan Turing: Life and Legacy of a Great Thinker* (Teuscher, C. A., ed.). Springer, New York (2004) 469–98.

Székely, E. Sur le problême de Tammes. *Ann. Univ. Sci. Budap. Roland Eötvös Nominatae Sect. Math.* 17 (1974) 157–75.

Tait, P. G. *Introduction to Quaternions.* Cambridge University Press (1873).

Takakura, H., Sato, A., Yamamoto, A. & Tsai, A. P. Crystal structure of a hexagonal phase and its relation to a quasicrystalline phase in Zn-Mg-Y alloy. *Phil. Mag. Lett.* 78 (1998) 263–70.

Takakura, H., Sato, T. J., Tsai, A. P., Sato, A. & Yamamoto, A. Crystal structure of hexagonal phases and its relation to icosahedral quasicrystalline phase in Zn–Mg–Re (Re = rare-earth) system. *Mat. Res. Soc. Symp. Proc.* 533 (1999) 129–34.

Tammes, P. M. L. On the number and arrangements of the places of exit on the surface of pollen-grains. *Recueil Travais Botaniques Néderlandais* 27 (1930) 1–84.

Tamura, N. The concept of crystalline approximants for decagonal and icosahedral quasicrystals. *Phil. Mag. A* 76 (1997) 337–56.

Tanaka, K., Yamabe, T. & Fukui K. (eds). *The Science and Technology of Carbon Nanotubes.* Elsevier (2000).

Tarnai, T. Symmetry of golf balls. *Katachi U Symmetry* (Ogawa, T., Miura, K., Masunari, T. & Nagy, D., eds). Springer, Tokyo (1996) 207–14.

Tarnai, T. Packing of equal circles in a circle. *Hyperspace* 7(2) (1998) 51–58.

Tarnai, T. Optimal packing of circles in a circle. In *Symmetry 2000* (Hargittai, I. & Laurent, T., eds). Portland Press, London (2000) 121–32.

Tarnai, T. & Gáspár, Zs. Multi-symmetric close packings of equal spheres on a spherical surface. *Acta Cryst. A* 43 (1987) 612–16.

Tenne, R., Margulis, L., Genut, M. & Hodes, G. Polyhedral and cylindrical structures of tungsten disulphide. *Nature* 360 (1992) 444–6.

Tenne, R., Homyonfer, M. & Feldman, Y. Nanoparticles of layered compounds with hollow cage-like structures (inorganic fullerene-like structures). *Chem. Mater.* 10 (1998) 3235–8.

Teo, B. K. & Zhang, H. Clusters of clusters: self-organization and self-similarity in the intermediate stages of cluster growth of Au–Ag superclusters. *Proc. Natl. Acad Sci. USA* 88 (1991) 5067–71.

Terrones, H. Nanomaterials with curvature. *Revista Especializada en Ciencias Quimico-Biológicas* 5 (2002) 47–56.

Terrones, H. & Mackay, A. L. The geometry of hypothetical graphite structures. *Carbon* 30 (1992) 1251–60.

Terrones, M. & Terrones, H. The role of defects in graphitic structures. *Fullerene Sci. Technol.* 4 (1996) 517–33.

Terrones, H. & Terrones, M. Curved nanostructured materials. *N. J. Phys.* 5 (2003) 126.1–126.37.

Terrones H., Terrones, M., López-Urías, F., Rodríguez-Manzo, J. A. & Mackay, A. L. Shape and complexity at the atomic scale: the case of layered nanomaterials. *Phil. Trans. Roy. Soc. London A* 362 (2004) 1471–2962.

Thompson, W. (Lord Kelvin). On the division of space with minimum partitional area. *London, Edinburgh and Dublin, Phil. Mag. J.* 24 (1887) 503–14.

Thompson, W. (Lord Kelvin). On homogeneous division of space. *Proc. Roy. Soc. London* 55 (1894) 1–16.

Thompson, D. W. On the thirteen semi-regular solids of Archimedes, and on their development by the transformation of certain plane configurations, *Proc. Roy. Soc. London A* 107 (1925) 181–8.

Thompson, D. W. *On Growth and Form* (1st ed.). Cambridge University Press (1917); (2nd ed. (1942); unabridged reprint (1996).

Torquato, S. & Stillinger, F. H. Multiplicity of generation, selection and classification procedures for jammed hard particle packings. *Jour. Phys. Chem.* 105 (2001) 11849.

Torquato, S., Truskett, T. M. & Debenedetti, P. G. Is random close packing of spheres well defined? *Phys. Rev. Lett.* 84 (2000) 2064–7.

Townsend, S. J., Lenowsky, T. J., Muller, D. A., Nichols, C. S. & Elser, V. Negatively curved graphite sheet model of amorphous carbon. *Phys. Rev. Lett.* 69 (1992) 921–4.

Truchet, S. Memoir sur les combinaisons. *Memoires de l'Academie Royale des Sciences* (1704) 363–72.

Tsai, A. P., Inoue, A., Yokoyama Y. & Masumoto, T. Stable icosahedral Al–Pd–Mn and Al–Pd–Re alloys. *Mater. Trans. JIM* 31 (1990) 98–103.

Turing, A. M. The chemical basis of morphogenesis. *Phil. Trans. Roy. Soc. London B* 237 (1952) 37–72.

Turnbull, D. D. Kinetics of solidification of supercooled liquid mercury droplets. *J. Chem. Phys.* 20 (1952) 411–24.

Turnbull, D. Under what conditions can a glass be formed? *Contemp. Phys.* 10 (1969) 473–88.

Ulam, S. Patterns of growth of figures: mathematical aspects. In: *Module, Symmetry, Proportion* (Kepes, G., ed.). Studio Vista, London (1966).

van Iterson, G. *Mathematische und Microscopisch-Anatomische Studien über Blattstellungen.* Fischer, Jena (1970).

Venkataraman, G., Sahoo D. & Balakrishnan V. *Beyond the Crystalline State.* Springer-Verlag, Berlin (1989).

Villars P. & Calvert, L. D. *Pearson's Handbook of Crystallographic Data for Intermetallic Phases.* American Society of Metals, Metals Park, Ohio (1986).

Voderberg, H. Zur Zerlegung der Ebene in kongruente Bereiche in Form einer Spirale. *Jahresber. Deutsch Math. Verein* 47 (1937) 159–60.

von Schnering, H. G. & Nesper, R. How nature adapts chemical structure to curved surfaces. *Angew. Chem.* 26 (1987) 1058–80.

von Schnering, H. G. & Nesper, R. Nodal surfaces of Fourier series: fundamental invariants of structured matter. *Z. Phys. B: Condens. Mat.* 83 (1991) 407–12.

von Schnering, H. G., Oehme, M. & Rudolf, G. Three-dimensional periodic nodal surfaces which envelope the threefold and fourfold cubic rod packings. *Acta Chemica Scand.* 45 (1991) 873–6.

Voronoi, G. F. Nouvelles applications des paramètres contínus à la théorie des formes quadratiques. *J. reine angew. Math.* 133 (1908) 97–178.

Wachman, A., Burt, A. & Kleinmann, M. *Infinite Polyhedra.* Technion, Haifa (1974).

Walter, A. Polyèdres couplés et polytopes. *Hyperspace* 9 (3) (2000) 22–9.

Watson R. E. & Weinert, M. Transition-metals and their alloys. *Solid State Phys.* 56 (2001) 1–112.

Weaire D. & Phelan, R. A counterexample to Kelvin's conjecture on minimal surfaces. *Phil. Mag. Lett.* 69 (1994) 107–10.

Weaire, D. & Phelan, R. A counter-example to Kelvin's conjecture on minimal surfaces. *Forma* 11 (1996) 209–13.

Weaire, D. & Rivier, N. Soap, cells and statistics – random patterns in two dimensions. *Contemp. Phys.* 25 (1984) 59–99.

Wefelmeier, W. Ein Geometrisches Modell des Atomkerns. *Z. Phys.* 107 (1937) 332–46.

Weierstrass, K, Untersuchungen über die Flächen, deren mittlere Krümmung überall gleich Null ist. *Monatsber. d. Berliner Akad.* (1866) 612.

Wells, A. F. *The Third Dimension in Chemistry.* Oxford University Press (1962).

Wells, A. F. *Three Dimensional Nets and Polyhedra.* Wiley, New York (1977).

Wells, A. F. Six new three-dimensional 3-connected nets. *Acta Cryst.* B39 (1983) 652–4.

Wenninger, M. J. *Polyhedron Models.* Cambridge University Press (1971).

Wenninger, M. J. *Dual Models.* Cambridge University Press (1983).

Weyl. H. S*ymmetry.* Princeton University Press (1952).

Whyte, L. L., Wilson, A. G. & Wilson, D. *Hierarchical Structures.* Elsevier, New York (1969).

Williams, R. *The Geometrical Foundations of Natural Structure.* Dover, London (1979).

Wills, J. M. A quasicrystalline sphere-packing with unexpected high density. *J. Phys. France* 51 (1990) 860–4.

Wilson, C. G., Thomas, D. K. & Spooner, S. J. The crystal structure of Zr_4Al_3. *Acta Cryst.* 13 (1960) 56–7.

Wohlgemuth, M., Yufa, N., Hoffman, J. & Thomas, E. L. Triply periodic bicontinuous cubic microdomain morphologies by symmetries. *Macromolecules* 34 (2001) 6083–9.

Woods, H. J. The geometrical basis of pattern design. V. *J. Text. Inst.* 27 (1936) 305–20.

Yang, Q.-B. & Andersson, S. Application of coincidence site lattices for crystal structure description: part I: $\Sigma = 3$. *Acta Cryst.* B 43 (1987) 1–14.

Zhang, Z., Ye, H. Q. & Kuo, K. H. A new icosahedral phase with m35 symmetry. *Phil. Mag. A* 52 (1985) L49–52.

Index

Page numbers in italics refer to illustrations